*Principles of Quantum Chemistry*

# Principles of
# Quantum Chemistry

*David V. George*

Dean of Studies
Notre Dame University of Nelson

**PERGAMON PRESS INC.**

New York · Toronto · Oxford · Sydney · Braunschweig

PERGAMON PRESS INC.
Maxwell House, Fairview Park, Elmsford, N.Y. 10523

PERGAMON OF CANADA LTD.
207 Queen's Quay West, Toronto 117, Ontario

PERGAMON PRESS LTD.
Headington Hill Hall, Oxford

PERGAMON PRESS (AUST.) PTY. LTD.
Rushcutters Bay, Sydney, N.S.W.

VIEWEG & SOHN GmbH
Burgplatz 1, Braunschweig

Printed in the United States of America
08 016925 2

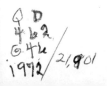

*To my parents*

# Contents

# Preface

Quantum chemistry has for many years been a very necessary part of the undergraduate chemist's education. In recent years the increasing correlation of theoretical and experimental results and the greater need for the experimentalist to understand the theory related to his work have made a good basic course in quantum chemistry at the undergraduate level a most urgent need. Though there are several excellent texts on quantum chemistry on the market, the majority do not include the sort of details the undergraduate chemist needs in order to read a text easily. In this text I have included more mathematical detail than usual: I hope this will make it easier for the student to use; and I especially hope that this will make it more unlikely that the student will lose sight of the development of the theory when he is wrestling with the mathematics that is intimately connected to the theory.

The background knowledge assumed for an understanding of the contents of this text is that knowledge of chemistry, physics and mathematics that any advanced junior or senior level chemistry student is quite familiar with. Everything else that is needed is included in the text. In particular, all mathematical methods used are developed from this basic assumed knowledge to the point where the student can fully understand the applications in the physical theory; and this is done at the places where the methods are first needed. It is the author's experience that most chemistry students are reluctant to think profoundly about the mathematical techniques unless they can see the immediate use of them: integration of the mathematics with the chemistry should make the former more palatable, so that the latter becomes more understandable.

This book will serve as a basic text for a course in quantum chemistry at the senior level. It is not meant to give a complete coverage of the field. Rather, it introduces the student to that part of quantum theory that is of major interest to chemists; it also introduces him to the most important use of quantum theory in chemistry — application to the understanding of chemical bonding. Two of the more important specific problems in chemical bonding — bonding in complexes and in conjugated organic molecules — are very briefly introduced; the very basic theory of spectroscopy is also briefly treated. With this latter material, and the basic theory developed in the main part of the text, the student should feel happy about entering more specialized undergraduate and graduate courses in quantum chemistry.

I should like to express my sincere appreciation to my colleague, Dr. D. Saraswati, who read the manuscript and offered many helpful suggestions for modifying it; I am also indebted to him for many useful discussions regarding its content. The manuscript was typed by my wife, Eila, who gave up a summer of her time to do it — I am most grateful to her.

*Nelson, British Columbia*                                    D. V. GEORGE

# 1

# Introduction

According to one of our most illustrious mentors, Albert Einstein, "the object of all science is to coordinate our experiences and to bring them into a logical system." Where does quantum theory fit into this description of science? As a chemist, you ask: Why should I study quantum chemistry? Let us try to answer these questions in a few sentences.

Classical mechanics evolved from Newton's laws. This mechanics adequately explained experimental results in the macro-world. It did, in fact, coordinate the experiences of the people of the time. On the macro-scale it is still adequate, but on the atomic and sub-atomic scales classical mechanics fails completely—it no longer coordinates the experiences of our experimental investigations. To understand the behavior of atomic and sub-atomic sized particles we need a new mechanics. Quantum mechanics seems to do the job of bringing our experiences with these particles into a logical system—this is why we who are scientists have faith in it. This is not to imply that we must accept this mechanics without reservation. We accept it so long as it explains our experimental results. It does so in all cases where we can be sure of the logical validity of the theoretical results. Now, the tests of a theory are that it explains our experiences with the physical world and predicts new ones. Until we can fail quantum theory on one of these counts we are justified in using it.

Now for the second question. This we can answer even more briefly. As chemists we are interested in the nature and interactions of matter. If matter is composed of atoms—and we have very good reason for supposing that it is—then every explanation we give in chemistry will ultimately depend on the laws governing the behavior of these atoms and their constituent parts, electrons and nuclei — this is where quantum mechanics comes in.

Before we look at the theory there are two subjects we should discuss very briefly — the use of models in the sciences and the use of mathematics in the sciences. In quantum chemistry we often use models to help us to understand the mathematical results of the theory. Therefore, it is important that we know exactly what we are doing when we use models. It is also important to appreciate the part that mathematics plays in the sciences in general and in quantum mechanics in particular. Quantum theory is essentially mathematical; consequently, if we are to fully understand it, we must understand its mathematical structure.

## 1-1   THE USE OF MODELS IN THE SCIENCES

Models help us to understand scientific phenomena. What is more they sometimes help us to predict new phenomena that can subsequently be tested by experiment. When we use a model, what we are doing is making an analogy between something unfamiliar and something familiar. When, for example, we compare gas molecules to billiard balls we use the familiar properties of the billiard balls to understand the unfamiliar properties of the gas molecules.

But there is a problem with the use of models: a model is only as good as the exactness of the analogy, and it is obvious that no model can be exactly analogous in all respects to the thing it is supposed to represent. If it were, we would be dealing with the real thing. The lesson here is that we must be careful not to carry our analogies too far. Because our model acts in a certain way does not necessarily mean that the thing it represents acts in the same way. The use of models for prediction, then, is that they predict possible phenomena; often it is more difficult, if not impossible, to imagine these phenomena without the models.

Here is an example to make this discussion more real. We consider the possibility of using a model to represent the electron. (This, of course, is an example with considerable relevance to our study of quantum chemistry.) In some of its properties that can be observed experimentally an electron acts like a particle. Since we cannot see electrons, let us consider them to be like particles that are familiar to us: we shall predict the properties of electrons by examining the properties of familiar particles, like tennis balls or bullets. If we do this we shall, in fact, be partially successful. But this model is not entirely satisfactory. In some experiments we discover that electrons act like waves. Perhaps, then, we should predict their properties by using water waves as our model. But then, what about the particle properties the electrons undoubtedly possess? We must use both models if we are to explain the properties of electrons. We make as much practical use as we can of each; but at the same time we must recognize the limitations of each. In fact, in a case such as this, the sort of intuitive reasoning that allows us to translate the properties of models to the system under consideration is not sufficient. Though the models are certainly useful for understanding the properties of electrons, we can understand these properties more fully only by resorting to a more basic and less intuitive type of reasoning. We shall not pursue this latter subject here since a major part of this book is concerned with this very point.

## 1-2  MATHEMATICS IN THE SCIENCES

So far as we who are scientists are concerned, mathematics is a tool. We use its symbolism and logical structure to aid us in formulating our theories; we use its concise methods to generalize the results of our experiments and experiences. The pure mathematician treats his subject as an intellectual discipline in which the most rigorous logic is essential — nothing less satisfies him; from a set of definitions or axioms he works by strictly logical steps toward his conclusions. Obviously, if the scientist is to make the maximum use of mathematics, he must aim for the same level of sophistication as the mathematician does. However, in a book of this type and at this level, insistence on complete mathematical rigor would often involve us in very lengthy mathematical digressions. We shall therefore insist on mathematical rigor so long as this does not hold up our progress to too great an extent; yet in all cases when we use mathematics with something less than complete rigor we shall make sure that we use it with at least "intuitive rigor."

One further point that we should make about the use of mathematics in the sciences is that mathematics to the scientist is a language: the symbols are the words of the language; the logical structure is the grammar. Now language is for communicating ideas — the language of mathematics, when used in the sciences, helps us to communicate scientific ideas to each other — and, indeed, to ourselves, for we have to coordinate our ideas before we can formulate them into theories: the conciseness of the mathematical language helps us here. But there is a problem with using the mathematical language: we first have to translate from our sensory world into the abstract world of mathematics. When we have used the mathematics to get results, we have to translate back to the sensory world for these results to have meaning. These translations have inherent dangers for the unwary—indeed, even for the wary—because it is often difficult to translate exactly from the physical world to the mathematical world. We can very easily understand this if we use the analogy of translating from one language to another—often we come across words that have no exact translation.

Why have we been talking about mathematics in a book on quantum chemistry? It is because quantum theory is essentially mathematical. The language and logic of quantum theory are mathematical. The results we obtain are mathematical; of course, we have to try to understand these results in terms of what exists in the physical world, but the original results are indeed mathematical ones.

## 1-3 SUMMARY OF THE BOOK AND HOW TO STUDY IT

Before you attack the main part of this book, some suggestions for studying quantum chemistry are in order.

Though quantum theory is essentially mathematical, there are physical models that can help us to understand it; but these are, in a way, adjuncts to the mathematical theory. Now, because of the possibility of using physical models we can at least partially understand quantum theory with little knowledge of mathematics. However, one thing is certain: we cannot apply the theory to problems of scientific interest without a good basic knowledge of the mathematics involved. What is more, we cannot really appreciate the significance of the theory in chemistry unless we understand the mathematics of quantum chemistry. And finally, we cannot appreciate the beauty of the logical structure of quantum theory without a very good knowledge of its mathematics. Therefore, in this book no attempt has been made to reduce to any great extent the amount of mathematics. However, mathematical digressions are distinguished by being printed in small type. This is not so that you can leave them out—it is to separate the basic mathematical methods from the applications of the methods; it will also serve to make these sections easy to identify for reference purposes. If you find some of the mathematical digressions and lengthy proofs tough going, skip them on first reading, but do go back and work through them until you fully understand them. It is much easier to appreciate the meaning of quantum mechanical results if you understand exactly how they were obtained.

Perhaps these words have discouraged you who have never been highly enthusiastic about mathematics. Do not let them. In this book the only knowledge of mathematics assumed is that which you almost certainly have and are reasonably familiar with if you have got as far as picking the book up. All other mathematical methods used are developed fully, either from scratch or from this knowledge you are assumed to have. So that you will know exactly where you stand, here is a list of what you are expected to know:

1. Trigonometry.
2. A very basic knowledge of analytic geometry.
3. Differentiation and integration of simple functions.
4. Multiple integration.
5. A knowledge of what partial differentiation involves.
6. A knowledge of how to solve simple, ordinary differential equations.
7. You should know what a vector is and some very simple properties of vectors.
8. A knowledge of the very basic concepts of the algebra of complex numbers and complex functions.

That's all. With this mathematical knowledge, which you will have covered in your early mathematics courses, and your basic knowledge of chemistry and physics, you can fully understand everything in this book. It would be less than honest to try to convince you that quantum mechanics is easy to understand. It is not. But it is within the grasp of the under-graduate chemist if he is prepared to think profoundly about the unfamiliar ideas introduced by the theory. When you remember that chemistry is probably ultimately understandable in terms of quantum theory, you will agree that it is more than well worth your while to make the effort needed to understand this theory.

Now, a few words about the structure of the book:

After a very brief review of the evolution of the physical sciences to the time of the introduction of quantum theory, we approach the problem of applying this theory to the understanding of the physical world. We do this at first in a way that makes it reasonably easy to accept the revolutionary ideas involved: we see how we can get at the Schroedinger equation, the basic equation of quantum chemistry.

In Chapter 3 we apply the Schroedinger equation to a very simple problem—the particle-in-a-box. This problem, though extremely idealized, gives us an insight into some of the basic features of quantum mechanics; it shows us how these features, such as quantum numbers and zero-point energy, result from the logic of the mathematical structure of the theory.

In Chapter 4 we start from the beginning again. We recognize that our first shot at the problem is not general enough for our purposes if we are to try to explain our complex chemical systems with quantum theory. We introduce a set of postulates. Some general consequences follow immediately from these postulates – we examine these in detail and try to relate our mathematical concepts to what we can sense in the physical world.

Chapter 5 deals entirely with angular momentum, a very important subject in quantum chemistry. Here we see how to deal with orbital angular momentum in quantum mechanical terms. Also, we find out how to bring into the theory the intrinsic angular momentum that some particles of chemical interest possess.

At this point we should have a basic understanding of the general form of the theory: we are ready to apply it to some further problems. Again the problems are simple and idealized. However, they illustrate some further features of quantum theory; also, they allow us to examine some of the mathematics of interest in the theory.

In Chapter 7 we consider the main approximation methods used to solve quantum mechanical problems. These are necessary, since most problems of practical interest cannot be solved exactly by quantum mechanical methods.

In Chapter 8 we look at a mathematical theory that is extremely useful in chemistry – group theory. Group theory allows us to take advantage of the symmetry properties that many of the systems we deal with possess. First, we examine the basic mathematics of the theory; then, we show how it can be applied to some simple chemical problems.

Many-electron atoms are the subject of Chapter 9. Systems of chemical interest are, of course, invariably many-electron systems. By examining some of the fundamental theory needed to deal with many-electron atoms, we can get some idea as to how to deal with many-electron systems in general.

By the end of Chapter 9 we are ready to consider one of the major problems in chemistry – the understanding of the chemical bond. First, we look at the two main theories of chemical bonding – molecular orbital theory and valence-bond theory. Then we devote a chapter each to two particular chemical bonding problems – the bonding in organic molecules and in transition metal complexes. These last two chapters serve as introductions to the theory of bonding in these compounds. They are not meant to be complete surveys of the present state of knowledge; rather, they prepare you for more specialized courses dealing specifically with these problems.

The last chapter considers another problem of considerable chemical interest—the understanding of spectra. Here again this is merely an introduction to the subject, a preparation for a course in molecular spectroscopy.

Well, that's it. Of course, we have not by any means exhausted the application of quantum theory to chemical problems. We have not, for example, even mentioned solid state theory or magnetic properties of molecules, two subjects of considerable interest to chemists. However, by the time you finish this book you should have sufficient understanding of the principles of quantum chemistry to feel happy about getting into more specialized courses in this most important field.

Just one more thing before you start the study of quantum theory: work through the exercises at the end of each chapter—they are an integral part of the text.

# 2

# The Early Development of Quantum Theory

## 2-1 REVIEW OF THE EARLIEST WORK IN QUANTUM THEORY

If we are to fully appreciate what quantum mechanics can do for us, we should have some knowledge of the work in the natural sciences that led up to its acceptance. We shall not detail this early work since you are probably reasonably familiar with it. However, mainly for completeness in this presentation, but also to refresh your memories, we shall briefly review the early experimental work and the theories that derived from this.

Classical physicists treated particles and waves as two entirely separate aspects of the physical world. This was only recognizing what their senses seemed to tell them. It is true that people had not always thought of material bodies as being made up of particles — certain Greek philosophers, for example, had proposed a continuous composition for matter. It is also true that people had not always thought of light as waves — three hundred years ago Newton supposed that light was a stream of particles. But, by the end of the nineteenth century, scientists were reasonably certain that matter was particles, and radiation such as light was waves — and that was that. It would be much simpler for us if things had stayed that way. They did not.

Toward the end of the nineteenth century it was discovered that electrons were emitted from a metal surface when light of high enough frequency fell on it (this is the photoelectric effect). This, in itself, was not a surprising observation. After all, water waves can wash pebbles from beaches, so why could not light waves wash electrons from metals? But it was not as simple as that. The maximum energy of photoelectrons was found to be independent of the intensity of the incident light; also, the electrons seemed to be emitted almost as soon as the light struck them. If light consisted of waves, the energy of the emitted electrons should have varied with the intensity of the light, and it would have taken a considerable time for any particular electron to pick up enough energy to escape from the metal. Another odd feature of the photoelectric effect was that the maximum energy of the photoelectrons depended on the frequency of the incident light.

In 1905, Einstein explained the photoelectric effect by using Planck's quantum theory. In 1900, Max Planck, a German theoretical physicist, had explained the nature of the radiation emitted by hot bodies by assuming it to come off as bursts of energy: he called these packets of energy quanta (we usually call them photons). The result of Planck's explanation was the equation

$$E = h\nu \tag{2-1}$$

$\nu$ is the frequency of the radiation; $E$ is the energy that a quantum of this frequency has; $h$ is now known as Planck's constant ($h = 6.63 \times 10^{-27}$ erg sec). Suppose that a minimum amount of energy is needed to release an electron from a metal, said Einstein: then either a quantum of light has this energy, or it does not — it is as simple as that. If the quantum's energy is too low, it cannot eject an electron; if it is more than the minimum needed, the electron is ejected with kinetic energy. Einstein gave us the equation for the process:

$$h\nu = w + T$$

$h\nu$ is the energy of the incident quantum of light; $w$ is the energy needed to eject an electron from the metal; $T$ is the energy of the ejected electron.

Many years later, in 1923, an American physicist, Arthur H. Compton, gave a striking experimental demonstration of the quantum theory of radiation. He showed that when X-rays were scattered by electrons the collisions between the quanta of the radiations and the electrons obeyed the laws of conservation of energy and momentum — this was further evidence for the particle nature of electromagnetic radiations.

So light consisted of particles. But it consisted of waves did it not? How could diffraction and interference, and other wave-like properties, be explained if it did not? Here was a problem.

The situation with matter at the turn of the century was very much happier. Certainly, this consisted of particles — the atoms of Dalton's theory. Not much was known about these atoms; but they were indeed particles – that much was certain. Naturally, scientists tried to discover the nature of atoms. The progress was slow, but in 1911, the New Zealand physicist, Rutherford, thought up a model for the atom to explain his experimental results. The atom, he said, consists of a comparatively very small nucleus with the electrons occupying most of its volume; nearly all the mass and all the positive charge is concentrated in the nucleus, which is only about $10^{-12}$ to $10^{-13}$ cm in diameter; the diameter of the atom is about $10^{-8}$ cm.

The next question to answer was: What were the electrons doing in the atom? They could not be static, for classical theory insisted that if they were they would fall into the nucleus. But they could not be moving about the nucleus either, because this, according to classical theory, would cause them to radiate energy – the result would be that eventually they would spiral into the nucleus. Here was a dilemma. Yet the Rutherford atom explained so much. Surely it was not to be rejected.

In 1913, Niels Bohr, a Danish physicist, made a revolutionary suggestion to solve the atomic problem. Let us suppose, he said, that there are certain allowed orbits in the atom in which the electrons have fixed amounts of energy; whilst in these orbits the electrons do not radiate energy, and the atom is therefore stable. This was a quantum condition imposed on a classical model. Admittedly, the imposition was entirely arbitrary; but it did seem to explain many of the experimental results that had worried scientists for some time. In particular, it explained the work of the spectroscopists.

Experimental spectroscopy had interested scientists for a long time. In the mid-seventeenth century, Newton had carried out experiments on the refraction of light by a glass prism; he deduced that white light was split by the prism into a spectrum of its component colored lights. However, little progress was made in spectroscopy until the end of the eighteenth and the beginning of the nineteenth centuries. About this time a notable contribution was made by a German, Josef Fraunhofer. Fraunhofer did some very significant experimental work on the sun's spectrum, and his accurate spectroscopic measurements led the way for experimental spectroscopy to become an exact science. In the three quarters of a century following Fraunhofer's work very accurate spectroscopic measurements were made by such men as Bunsen, Kirchhoff and Angstrom. In 1885, J. J. Balmer discovered that in the atomic hydrogen spectrum there was a relationship between the frequencies of the lines in the visible region, given by

$$\bar{\nu} = R\left(\frac{1}{2^2} - \frac{1}{n_1{}^2}\right) \tag{2-2}$$

$\bar{\nu}$ are the wave numbers of the spectral lines; $R$ is a constant known as the *Rydberg constant*; $n_1$ is an integer greater than two, i.e., $n_1 = 3, 4, 5, \ldots$. Later it was discovered that this relationship, which gave a series of lines in the visible region of the spectrum, could be generalized to account for the occurrence of other series of lines in different parts of the spectrum. The more general equation is

$$\bar{\nu} = R\left(\frac{1}{n_2{}^2} - \frac{1}{n_1{}^2}\right) \tag{2-3}$$

The various series of lines in the hydrogen atom spectrum are understandable in terms of this equation. Figure 2-1 explains these series on the basis of the allowed energy levels for hydrogen. The Lyman series

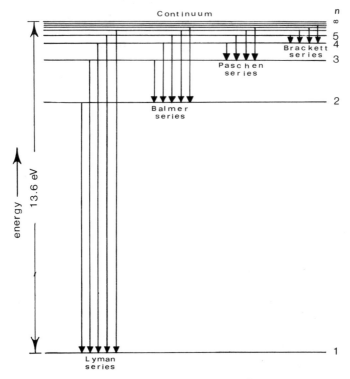

**Fig. 2-1**   Energy level diagram for the hydrogen atom illustrating how the various spectral series are formed.

results from putting $n_2 = 1$ and $n_1 = 2, 3, 4, \ldots$; the Balmer series is obtained by putting $n_2 = 2$ and $n_1 = 3, 4, 5, \ldots$; and so on.

Though a very considerable amount of accurate experimental spectroscopic results were available, at the time when Bohr made his revolutionary proposal not much progress had been made in understanding the theoretical basis for them. Bohr's model of the atom was able to explain the observable hydrogen spectrum very nicely. According to this model, the discrete lines in the spectrum were due to electrons jumping from one allowed orbit to another; energy was emitted when this happened, but this energy could correspond only to the difference in energy between two allowed orbits. And not only was the Bohr theory able to explain the spectrum of the hydrogen atom in a qualitative way — it gave the Rydberg constant with quite surprising accuracy. However, everything was not yet completely satisfactory: Bohr's theory could not explain the spectra of atoms with more than one electron; nor could it explain the fine structure that could be observed in the spectral lines.

In the years following the introduction of the simple Bohr model, some theoretical work was done to refine it. W. Wilson and A. Sommerfeld independently modified the theory by allowing elliptical orbits for the electrons — this was partially successful in explaining the fine structure in atomic spectra. However, the clue to a more satisfactory solution of the problem came in 1924, when Louis de Broglie, a French physicist, proposed that matter possesses wave-like properties. According to de Broglie, a particle with momentum $p$ has a wavelength $\lambda$ associated with it; $\lambda$ is given by the equation

$$\lambda = \frac{h}{p} \tag{2-4}$$

By using this equation, one could get a direct relationship between the particle properties as expressed by momentum and the wave-like properties as expressed by the wavelength.

Soon the wave nature of matter was demonstrated experimentally. In 1927, C. Davisson and L. H. Germer, working in the United States, showed that electrons could be diffracted, just like light waves and other electromagnetic radiations. These workers bombarded a nickel crystal with a beam of low velocity electrons and examined the reflected beam of electrons. The diffraction pattern they observed was very similar to that obtained by X-ray diffraction experiments. This was convincing evidence for the wave nature of matter.

So scientists were left with a very perplexing problem: Not only did light and other electromagnetic radiations act like particles under certain conditions, but particles acted like waves. In our macroscopic world particles and radiations are entirely different aspects of nature; in the atomic world the distinction appeared to be much more nebulous.

About this time, in 1927 to be exact, Werner Heisenberg, a German mathematician, proposed his very famous *uncertainty principle*. Classical mechanics had always assumed that it was possible to measure simultaneously and precisely such properties of particles as momentum and position. This simultaneous knowledge of position and momentum for some instant in time then allowed prediction of these properties at any other time. According to the uncertainty principle, it is not possible to measure simultaneously the position and momentum of a particle. By endowing particles with a dual wave-particle nature, this contradiction to classical mechanical ideas was understandable. If a particle had some of the properties of waves, an uncertainty principle was a logical consequence. We shall not pursue this very important aspect of quantum theory here since we consider it in more detail in Chapter 4.

It is at this point that we shall start our detailed presentation of quantum theory.

## 2-2  QUANTUM THEORY

An explanation of the dual nature of what were previously called particles and waves is to be found in quantum theory. We must reject classical mechanics when we deal with atomic and sub-atomic phenomena. We must devise a whole new mechanics to explain what goes on in the atomic world. We must replace the equations of classical mechanics with new ones, better able to explain what we can experimentally observe.

Can we derive these new equations from the old ones we used in classical mechanics? — classical mechanics, we remember, worked well enough in the macroscopic world. The answer is that we cannot. To build up a quantum mechanical theory we start again from the beginning. We make some postulates that at first sight seem rather arbitrary, and formulate the theory on the basis of these postulates. The question that immediately springs to mind is: What justification do we have for this approach? The justification is in the agreement we get when we try to understand our experimental results in terms of quantum theory. This, of course, is not conclusive proof that quantum theory is absolutely correct. Yet it explains so much that few scientists are even cautious about using it nowadays. After all, one of the major criteria for scientific truth is agreement with experiment. Let us just say that unless a better theory is discovered we had best accept the one we have, albeit with the reservations that, as scientists, we must always hold.

## 2-3 THE SCHROEDINGER EQUATION

We are about to start our detailed study of quantum mechanics. How shall we start? There are several lines of approach: Erwin Schroedinger, a Viennese physicist, used wave mechanical methods to formulate the theory; Werner Heisenberg used matrix methods for the same purpose; Paul Dirac, an English physicist, used an entirely symbolic method. All three methods give the same end results. However, Schroedinger's is the easiest to understand for students meeting quantum mechanics for the first time. This is because it allows us to make analogies with familiar vibrations like vibrating strings. We shall, therefore, use this method, though the others have some mathematical advantages. Fortunately, nearly all the problems we meet in chemistry can be understood in terms of Schroedinger's formulation.

The Schroedinger formulation[1], in common with the other formulations, requires some postulates on which to base the theory. The basic equation of this method is the so-called Schroedinger equation. We cannot derive this equation from classical mechanics without making nonclassical postulates. We can, therefore, if we wish, use this equation as a postulate and work from there. However, this is usually not a very acceptable method to students, so we shall not use it. Instead, we shall derive the Schroedinger equation from some postulates based on experimental evidence—the wave nature of particles. It is important to realize that we still have to make some postulates. But, the postulates we make here should be more acceptable to you than others we could make.

Though the method we follow here gives us a starting point, we shall discover later that it is not sufficiently general. Then we shall go back to the beginning, so to speak, and start again: we shall make a very general set of postulates and build up the theory from these. Why do we not immediately take the general approach? Well, the reason is that by doing it this way we shall be in a better position to understand and appreciate the general approach when we get to it, because by that time we should have some idea as to what quantum mechanics is all about. So let us get on with the derivation of the Schroedinger equation from postulates based on experimental evidence.

Particles show wave-like properties—this we can easily verify by experiment. The first thing we shall do, therefore, is very briefly review some aspects of classical wave motion. To do this we shall use the very familiar example of waves in a taut string fixed at both ends.

If we pluck the string, it will vibrate. However, the fact that the ends are fixed prevents it from taking up all waveforms. Here we shall be interested in the waveforms known as *standing waves*; some of the possible standing waves are shown in Fig. 2-2. If we analyze these waveforms,

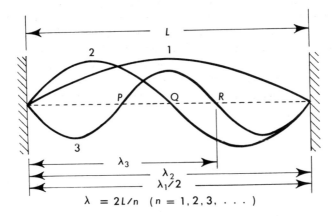

**Fig. 2-2**   Some of the standing waves that can result from plucking a taut string.

we see that most points on the string undergo periodic displacements, but that points such as $P$ and $Q$ remain stationary. The stationary points are called *nodes*, and the particular standing waves in which they occur are characterized by the number of nodes. Neglecting the end points, the number of nodes for the various standing waves are $0, 1, 2, 3, \ldots$, as shown in Fig. 2-2. If we wished to formulate a mathematical theory to describe the wave motion in the string, we would have to take into account the occurrence of the nodes. We would try to find a mathematical function that would describe the displacements of the various parts of the string, keeping in mind that the wavelength is restricted to certain values. From Fig. 2-2, we can readily see that the allowed wavelengths $\lambda$ are given by

$$\lambda = \frac{2L}{n}$$

$L$ is the length of the string; $n$ is any integer except zero. The important point to note is that the mathematics would have to bring these integers into the theory in a natural way. Remember that integers had occurred in the Bohr theory of the hydrogen atom (cf. Eqs. (2-2) and (2-3)) as a result of the rather arbitrary introduction of quantum restrictions. Schroedinger utilized their natural occurrence in classical wave theory and, by basing his development of quantum mechanics on wave mechanical methods, caused the mathematics to introduce them naturally. But before we apply wave mechanical reasoning to quantum mechanics we must examine the classical situation a little further.

We said a moment ago that if we wished to develop a mathematical theory for the wave motion of the string, we would try to find a mathematical function to describe the motion. We shall not do this here since it is a problem you will be quite familiar with from introductory physics and physical chemistry courses†. The motion, you will recall, can be described by a function $\Psi$ that gives the displacements of the various parts of the string; a convenient form for $\Psi$ for our present purposes is

$$\Psi = A \sin \frac{2\pi x}{\lambda} \cos 2\pi\nu t \qquad (2\text{-}5)$$

$A$ is a constant equal to the maximum height of the wave; $x$ is the distance from the origin if we place the origin at the left-hand end of the string; $\nu$ is the frequency of the vibrations; and $\lambda$, remember, is the wavelength. At some particular time $t$ this function takes the form

$$\Psi_m = A_m \sin \frac{2\pi x}{\lambda} \qquad (2\text{-}6)$$

$A_m$ again is the height of the wave. $\Psi_m$ now gives the instantaneous displacements of the various parts of the string; i.e., $\Psi_m$ describes what the wave would look like if we took a snapshot of it at some particular time $t$. The form of $\Psi_m$ is shown in Fig. 2-3.

We have been examining wave motion with a view to applying the ideas involved to quantum theory. We are now in a position to obtain an equation that will allow us to accomplish this objective. Equation (2-5) is a solution of an equation that we can obtain by differentiating Eq. (2-5) twice with respect to $x$; doing this we get

$$\frac{\partial^2 \Psi}{\partial x^2} = -\frac{4\pi^2}{\lambda^2} A \sin \frac{2\pi x}{\lambda} \cos 2\pi\nu t$$

---

†For a somewhat more general discussion of wave motion in strings you can consult, for example, Andrews, D. H., *Introductory Physical Chemistry*, McGraw-Hill, New York, 1970, p. 341ff.

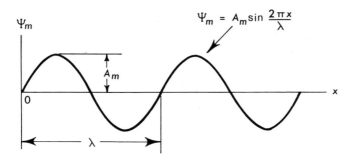

**Fig. 2-3**  Plot of a function suitable for describing the waveforms in Fig. 2-2 at some given instant in time.

or, using Eq. (2-5),

$$\frac{\partial^2 \Psi}{\partial x^2} + \frac{4\pi^2}{\lambda^2} \Psi = 0 \tag{2-7}$$

Before we make use of this equation we shall make a slight change in it. Most of our discussion of quantum mechanics will be concerned with time-independent systems. If $\Psi$ in Eq. (2-7) were time-independent, we could write an ordinary derivative instead of the partial derivative; Eq. (2-7) would then take the form

$$\frac{d^2\Psi}{dx^2} + \frac{4\pi^2}{\lambda^2} \Psi = 0 \tag{2-8}$$

It is this last equation that we shall apply to quantum mechanics. Equation (2-8) resulted from our consideration of a classical wave motion. But particles of matter show wave-like properties, so we shall assume that a similar equation is applicable to matter. Also, we recall that de Broglie gave an equation, Eq. (2-4), relating the wave nature of matter to its particle nature: it was $\lambda = h/p$. Substituting this into Eq. (2-8), we get

$$\frac{d^2\Psi}{dx^2} + \frac{4\pi^2 p^2}{h^2} \Psi = 0 \tag{2-9}$$

In chemistry, energy is usually of greater concern to us than momentum, so we try to bring the energy into the equation. Restricting ourselves to conservative systems, for which the total energy remains constant, the total energy $E$ is the sum of the kinetic energy $T$ and the potential energy $U$. If we consider a single particle of mass $m$, we then get the equation

$$E = T + U = \frac{mv^2}{2} + U = \frac{p^2}{2m} + U \tag{2-10}$$

i.e.,

$$p^2 = 2m(E - U) \tag{2-11}$$

Substitute Eq. (2-11) into Eq. (2-9); we get

$$\frac{d^2\Psi}{dx^2} + \frac{8\pi^2 m}{h^2}(E - U)\Psi = 0 \tag{2-12}$$

The quantity $h/2\pi$ appears so often in quantum mechanical equations that it is convenient to write $\hbar$ for it. Using this notation, we usually write Eq. (2-12)

$$\frac{d^2\Psi}{dx^2} + \frac{2m}{\hbar^2}(E - U)\Psi = 0 \tag{2-13}$$

Equation (2-13) is the *Schroedinger equation for a stationary state*. (The stationary state refers to a state of the particle in which the energy does not change with time. Later, we shall see that a more general equation is needed if this condition is not met.) We obtained it by applying wave mechanical methods to matter. Our justification for this approach is the experimental evidence for the wave nature of matter.

So far we have confined ourselves to one dimension. In reality our systems move in three dimensions. It is easy to generalize the Schroedinger equation to make it applicable to three dimensions. When we do this we get

$$\frac{\partial^2\Psi}{\partial x^2} + \frac{\partial^2\Psi}{\partial y^2} + \frac{\partial^2\Psi}{\partial z^2} + \frac{2m}{\hbar^2}(E - U)\Psi = 0 \tag{2-14}$$

## 2-4 INTERPRETATION OF $\Psi$

The problem now is to interpret the meaning of $\Psi$. After all, our system is supposed to have particle properties as well as the wave properties we used to obtain the Schroedinger equation. $\Psi$ for a wave gives its amplitude; but what can it mean for a particle with wave-like properties? This question was the subject of much discussion and argument when wave mechanics was first applied to matter. But before we examine the question let us decide what to call $\Psi$, since we shall be using it continually from now on.

It is conventional to call $\Psi$ a *wavefunction*. The reason for this is obvious: it originated in the application of wave mechanics to the understanding of matter. However, when using the term wavefunction, we should not forget that we are applying it to something that has a particle nature as well as a wave nature.

What about the interpretation of the wavefunction? Let us interpret it in terms of an example—the electron. It was Max Born[2], a German physicist, who suggested what is now taken to be the most satisfactory interpretation for the physical meaning of Ψ. Schroedinger interpreted it in terms of an electron smeared out into a cloud of charge: Ψ, he said, was a measure of the charge density in the cloud. This was a contradiction of what had always been accepted, for the electron and other units of matter had always been imagined to be precisely located in space—almost point particles. Born rejected the Schroedinger interpretation. He said that, for an electron, Ψ is in some way a measure of the probability of finding it. However, Ψ itself cannot represent this probability since this would restrict it to being a real function that is always positive or zero. In fact, Ψ can be negative, and can even be a complex function. We can get a clue to the interpretation of Ψ by recognizing that it is related to the amplitude of a wave in classical wave theory. In classical theory the energy of an electromagnetic field at a given point is proportional to the square of the amplitude of the wave at that point. We might therefore expect that the probability of finding the electron would be related to $\Psi^2$ if Ψ is a real function, or more generally to $\Psi^*\Psi$, allowing for Ψ to be a complex function. To be more precise, we postulate that the probability of finding the electron in the volume element $dxdydz$ is given by $\Psi^*\Psi dxdydz$.

There is one further point we should note before we leave the question of the interpretation of the wavefunction in terms of probabilities. In classical mechanics we often work with probabilistic descriptions of systems. However, we do so due to practical necessity—when we have huge numbers of particles (e.g., molecules) in a system, it is impractical to determine the initial states of all the particles in order to determine future or past states. We therefore sometimes resort to statistical descriptions of our classical mechanical systems. The probabilistic descriptions of quantum mechanical systems are of a very different type. Here we have to resort to the use of probabilities even when we are dealing with a single particle. In quantum mechanics it is the wavefunction that describes the state of a particle, and the physical interpretation of this is in terms of the probability of finding the particle in some particular region of space. So probabilistic descriptions of systems are inherent in quantum theory.

## 2-5  RESTRICTIONS ON THE WAVEFUNCTION

Now that we have given an interpretation to $\Psi$ we must impose some restrictions on it: it is a mathematical function, but it describes a physical system. Therefore, we insist that it should be a physically acceptable function.

Since $\Psi^*\Psi dxdydz$ represents the probability of finding the electron, the restrictions are that $\Psi$ must be everywhere finite, single-valued and be square integrable. The last restriction means that the probability of finding the electron in all space must be finite.

The probability of finding the electron somewhere in space is in fact unity. We can therefore write

$$\int_{-\infty}^{+\infty} \Psi^*\Psi dxdydz = 1 \qquad (2\text{-}15)$$

(Note here that the integral is really a multiple integral, though we have written it with a single integral sign. It is mathematically conventional, and very much more convenient, to write multiple integrals like this. How, then, do we know how many integral signs should be present? This is always obvious from the nature of the problem being studied.) When the condition Eq. (2-15) is imposed on the wavefunction, we say that it has been normalized to unity, or simply that it has been *normalized*.

Though when dealing with the restrictions on the wavefunctions we have considered only a single particle, we can easily extend the ideas to many-particle systems.

This is where we shall stop our development of quantum mechanics from the approach we have taken here. Admittedly, we have developed the theory very little, but it is sufficient to enable us to apply it to a very simple problem — we choose the familiar particle-in-a-box problem. But before we do this, let us summarize what we have done so far.

## SUMMARY

1. We very briefly reviewed the experimental work that led up to the acceptance of quantum theory by physicists; we reviewed the reasons for this acceptance and showed why it was thought that quantum theory could lead to an understanding of the physical world.

2. We mentioned that there are several ways of developing the theory of quantum mechanics. All require starting from some postulates. Here we chose a method that is acceptable to the person coming into contact

with the unfamiliar world of quantum mechanics for the first time. It should be acceptable because it allows us to obtain the fundamental equation of quantum mechanics using an analogy that is familiar — wave motion.

3. We started by briefly examining classical wave motion; then we brought in the experimental evidence for the wave nature of matter, as expressed by the de Broglie equation relating wavelength $\lambda$ associated with matter to its momentum $p$:

$$\lambda = \frac{h}{p} \tag{2-4}$$

4. Applying the classical wave equation to matter, we derived the *Schroedinger equation for a stationary state* for a single particle of mass $m$:

$$\frac{\partial^2 \Psi}{\partial x^2} + \frac{\partial^2 \Psi}{\partial y^2} + \frac{\partial^2 \Psi}{\partial z^2} + \frac{2m}{\hbar^2}(E - U)\Psi = 0 \tag{2-14}$$

5. $\Psi$ is called a *wavefunction*. It is a mathematical function representing the physical state of the system.

6. The probability of finding an electron (using an electron as an example of a physical system) in the volume element $dxdydz$ is given by $\Psi^*\Psi dxdydz$ — this is the physical interpretation of $\Psi$.

7. Because $\Psi$ represents the physical state of the electron, it must obey certain restrictions — it must be finite, single-valued and square integrable.

# 3

# The Particle-in-a-box

## 3-1  INTRODUCTION

In chemistry, to deal with a physical system in terms of quantum mechanics we invariably have to solve the Schroedinger equation for the system. Unfortunately, there are not many problems for which the Schroedinger equation can be solved exactly. Certainly, most of the problems of real practical interest involve Schroedinger equations that are way beyond our capabilities to solve. This does not mean, however, that we cannot deal with these problems at all. We can—but we have to use approximation methods. Later, we shall look at these approximation methods; presently, we shall consider a problem for which the Schroedinger equation can be solved. Admittedly we shall be studying an idealized system; also, the problem is very simple—almost naïvely simple. These facts, however, do not detract from the value of the results, because this system illustrates some of the most important features of quantum mechanics: it shows how quantization of energy, quantum numbers and zero-point energy arise naturally from the mathematics of the theory.

So let us tackle the problem. First we outline the nature of the problem; then we set up a Schroedinger equation for the system. We solve the equation. This gives us expressions for the wavefunction of the system and the energy. We briefly examine the meaning of the results.

## 3-2  THE PARTICLE-IN-A-BOX

We consider a particle enclosed by a box. (This is not an entirely hypothetical problem we are choosing: it approximates, for example, a gas molecule enclosed in a container.) Let the mass of the particle be $m$. The lengths of the edges of the box are $a$, $b$, and $c$ so that the volume of the box is $abc$. Inside the box the potential energy is assumed to be zero; everywhere outside the box the potential energy is assumed to be infinite. If we consider a rectangular coordinate system with the origin at the corner of the box, the system is represented by Fig. 3-1.

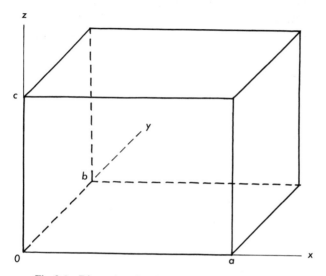

**Fig. 3-1**   Dimensions for the particle-in-a-box problem.

The Schroedinger equation for the system is (cf. Eq. (2-14))

$$\frac{\partial^2 \Psi}{\partial x^2} + \frac{\partial^2 \Psi}{\partial y^2} + \frac{\partial^2 \Psi}{\partial z^2} + \frac{2m}{\hbar^2}(E - U)\Psi = 0$$

We shall first consider the region outside the box where $U = \infty$. In this region the equation becomes

$$\frac{\partial^2 \Psi}{\partial x^2} + \frac{\partial^2 \Psi}{\partial y^2} + \frac{\partial^2 \Psi}{\partial z^2} + \frac{2m}{\hbar^2}(E - \infty)\Psi = 0 \tag{3-1}$$

or, since $E$ is obviously very small compared to infinity,

$$\frac{\partial^2 \Psi}{\partial x^2} + \frac{\partial^2 \Psi}{\partial y^2} + \frac{\partial^2 \Psi}{\partial z^2} = \infty \Psi \tag{3-2}$$

The solution outside the box is, therefore, $\Psi = 0$ because there is no function that is finite, single-valued and square integrable that satisfies Eq. (3-2). This means that $\Psi^2 = 0$ in this region; consequently, there is zero probability of finding the particle outside the box—this is what we would expect. Furthermore, if we are to get a physically reasonable solution, the function for the region outside the box must join smoothly with the function for the region inside the box. Therefore, since $\Psi = 0$ outside the box, it must approach zero at the boundaries inside the box.

We can now consider the region inside the box. We assumed that the potential energy is zero in this region, so the Schroedinger equation becomes

$$\frac{\partial^2 \Psi}{\partial x^2} + \frac{\partial^2 \Psi}{\partial y^2} + \frac{\partial^2 \Psi}{\partial z^2} + \frac{2mE}{\hbar^2} \Psi = 0 \qquad (3\text{-}3)$$

This is a partial differential equation because the first three terms are partial derivatives. We can solve it by a method that is often applicable to simple, partial differential equations. The method is called *separation of variables*. Unfortunately, we can use this method for only a few types of equations that occur in quantum mechanics. In quantum chemistry—and this is the section of quantum mechanics that concerns us—most problems of interest are not susceptible to solution by separation of variables—this is one of the major reasons why we are not able to rigorously solve the equations we get in quantum chemistry. However, despite this the method is so useful for solving simple problems of quantum mechanics—problems that illustrate very well some of the main features of the theory—that we shall digress to explain it before continuing. Furthermore, we make frequent reference to the method in future chapters, so you should take careful note of it now.

## Solution of Partial Differential Equations by Separation of Variables

The trick in this mathematical technique is to reduce a partial differential equation to a system of ordinary differential equations: the ordinary differential equations are then solved by usual methods.

We could illustrate the method by solving the problem at hand. Let us instead choose an easier example, because by so doing we shall not lose sight of the method in the complexity of the mathematics. We consider the equation

$$\frac{\partial u}{\partial x} - \frac{\partial u}{\partial y} = 0 \qquad (3\text{-}4)$$

To solve this equation we assume a solution of the form

$$u(x, y) = X(x)Y(y) \qquad (3\text{-}5)$$

Here, $X$ is a function of $x$ only; $Y$ is a function of $y$ only. This seems a very arbitrary assumption to make; why can we do this? The answer is: Because it solves the problem—we are being wise after knowing the solution. In other words, if we suspect that a partial differential equation may be solved by separation of variables, we try a solution of the form Eq. (3-5) to see if it works.

Substitute Eq. (3-5) into Eq. (3-4); we get

$$Y\frac{dX}{dx} - X\frac{dY}{dy} = 0 \qquad (3\text{-}6)$$

Note that we can now write ordinary derivatives. We can do this because $X$ is a function of $x$ only and $Y$ is a function of $y$ only. Now divide Eq. (3-6) by $U = XY$; we get

$$\frac{1}{X}\frac{dX}{dx} - \frac{1}{Y}\frac{dY}{dy} = 0 \qquad (3\text{-}7)$$

This is an identity, which means that it must be true for all possible values of $x$ and $y$. Therefore,

$$\frac{1}{X}\frac{dX}{dx} = \frac{1}{Y}\frac{dY}{dy} = c \qquad (3\text{-}8)$$

$c$ is an arbitrary constant.

Why is Eq. (3-8) true? We have defined $X$ as being a function of $x$ only; so the first term in Eq. (3-7) is a function of $x$ only. Similarly, the second term is a function of $y$ only. Now suppose we fix the value of $x$: the first term becomes a numerical constant. But the equation must be true for all values of $y$. Therefore, the second term also must be equal to this constant. On the other hand, we could fix $y$ and vary $x$; again, we would arrive at the conclusion that Eq. (3-8) is true.

What have we gained by doing all this? We have separated our two variables into two equations, each equation involving only one of the variables. Let us write the equations again:

$$\frac{1}{X}\frac{dX}{dx} = c \qquad \frac{1}{Y}\frac{dY}{dy} = c$$

These are ordinary differential equations. They can be solved by the usual methods. In this example the solutions are

$$X = Ae^{cx} \qquad Y = Be^{cy} \qquad (3\text{-}9)$$

$A$ and $B$ are arbitrary constants; they can be combined into one constant $D$ to give the solution

$$u(x, y) = X(x)\,Y(y) = ABe^{c(x+y)} = De^{c(x+y)} \qquad (3\text{-}10)$$

We now apply what we have learned to the equation that provoked the discussion of the separation of variables method of solving partial differential equations.

The equation we are trying to solve is

$$\frac{\partial^2 \Psi}{\partial x^2} + \frac{\partial^2 \Psi}{\partial y^2} + \frac{\partial^2 \Psi}{\partial z^2} + \frac{2mE}{\hbar^2}\Psi = 0$$

We try the substitution

$$\Psi = X(x)Y(y)Z(z) \qquad (3\text{-}11)$$

This gives

$$Y(y)Z(z)\frac{d^2X}{dx^2}+X(x)Z(z)\frac{d^2Y}{dy^2}+X(x)Y(y)\frac{d^2Z}{dz^2}+\frac{2mE}{\hbar^2}X(x)Y(y)Z(z)=0$$

$$(3\text{-}12)$$

Proceeding as we did in the illustrative example, we divide through by $\Psi = X(x)Y(y)Z(z)$; we get

$$\frac{1}{X}\frac{d^2X}{dx^2}+\frac{1}{Y}\frac{d^2Y}{dy^2}+\frac{1}{Z}\frac{d^2Z}{dz^2}+\frac{2mE}{\hbar^2}=0 \qquad (3\text{-}13)$$

This is an identity, true for all values of $x$, $y$, and $z$, so we can separate it into three equations. The equations are

$$\frac{d^2X}{dx^2}+\frac{2mE_x}{\hbar^2}X=0 \qquad (3\text{-}14\text{a})$$

$$\frac{d^2Y}{dy^2}+\frac{2mE_y}{\hbar^2}Y=0 \qquad (3\text{-}14\text{b})$$

$$\frac{d^2Z}{dz^2}+\frac{2mE_z}{\hbar^2}Z=0 \qquad (3\text{-}14\text{c})$$

Note that to satisfy the identity we have broken down the constant term $2mE/\hbar^2$ by writing $E$ as

$$E=E_x+E_y+E_z \qquad (3\text{-}15)$$

We shall solve one of the Eqs. (3-14). Naturally, because of the symmetry of the problem, the solutions to the others will be similar.

Let us consider Eq. (3-14a). This is a second-order, linear differential equation with constant coefficients. You should be familiar with equations of this type. The solution can be seen by inspection to be

$$X=A\exp\left(\frac{ix\sqrt{2mE_x}}{\hbar}\right)+B\exp\left(-\frac{ix\sqrt{2mE_x}}{\hbar}\right) \qquad (3\text{-}16)$$

where $A$ and $B$ are arbitrary constants. We can convert this to a more convenient form for our purposes by using the substitutions

$$\exp\left(\frac{ix\sqrt{2mE_x}}{\hbar}\right)=\cos\left(\frac{\sqrt{2mE_x}}{\hbar}x\right)+i\sin\left(\frac{\sqrt{2mE_x}}{\hbar}x\right)$$

$$\exp\left(-\frac{ix\sqrt{2mE_x}}{\hbar}\right)=\cos\left(\frac{\sqrt{2mE_x}}{\hbar}x\right)-i\sin\left(\frac{\sqrt{2mE_x}}{\hbar}x\right)$$

We get

$$X = C \cos\left(\frac{\sqrt{2mE_x}}{\hbar}x\right) + D \sin\left(\frac{\sqrt{2mE_x}}{\hbar}x\right) \qquad (3\text{-}17)$$

$C$ and $D$ are new arbitrary constants, replacing $A$ and $B$.

We would now like to evaluate the constants $C$ and $D$. This we can do by examining the boundary conditions. To determine $C$ we recognize that $X = 0$ at $x = 0$. Therefore, $C = 0$. $X$ is also zero at the other end of the box, i.e., where $x = a$. We can satisfy this condition if

$$E_x = \frac{n_x^2 \pi^2 \hbar^2}{2ma^2} \qquad (3\text{-}18)$$

and $n_x$ is an integer, i.e., $n_x = 1, 2, 3, \ldots$ ($n_x$ cannot be zero, for reasons we examine in a moment); $n_x$ is called a *quantum number*.

The results we have just obtained can be understood more easily if we present them in diagrammatic form; this is done in Fig. 3-2. (Remember

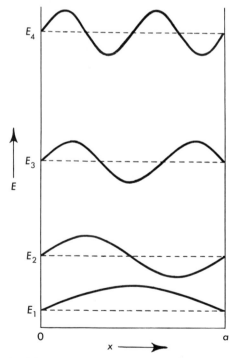

**Fig. 3-2**  Energy levels with corresponding wavefunctions for a particle in a one-dimensional box of length $a$.

that we are still working in one dimension.) There will be an infinite number of energy levels with corresponding wavefunctions – we have shown only the first four.

Before we determine the constant $D$ in Eq. (3-17) we shall make a preliminary examination of the results because they show two very important quantum mechanical effects. Later in this chapter we shall make a further examination of the results of the particle-in-a-box problem.

The first effect is that the energy is quantized. Examining the mathematics, we see that the restriction of the energy to certain values has been brought about by the imposition of boundary conditions. It is the fact that we have restricted the particle within the boundaries of the box that has caused its energy to be quantized.

The second important effect to emerge from these results is the concept of zero-point energy. Rewriting Eq. (3-17) with Eq. (3-18) for $E_x$ substituted into it, we get (remembering that $C = 0$)

$$X = D \sin \left( \frac{n_x \pi x}{a} \right) \tag{3-19}$$

If our particle is to be in the box, then $X$ cannot be zero everywhere inside the box; i.e., $n_x$ cannot be zero. This means that the lowest energy the particle can have is not zero. If it were confined to one dimension, the lowest value for its energy would be given by substituting $n_x = 1$ into Eq. (3-18); doing this, we get

$$E_x = \frac{\pi^2 \hbar^2}{2ma^2} \tag{3-20}$$

This, then, would be the so-called *zero-point energy*.

We have still not evaluated the second constant, $D$, in Eq. (3-17). How do we do this? We remember that the wavefunction must be normalized (cf. Eq. (2-15)):

$$\int_0^a X^2 dx = 1 \tag{3-21}$$

The limits in the general normalization expression are $-\infty$ and $+\infty$; the limits in Eq. (3-21) are zero and $a$ because the particle is only allowed to be between zero and $a$ in our coordinate system. Now substitute Eq. (3-19) into Eq. (3-21):

$$D^2 \int_0^a \sin^2 \left( \frac{n_x \pi x}{a} \right) dx = 1 \tag{3-22}$$

This is a standard integral; it gives

$$D = \sqrt{\frac{2}{a}}$$

Finally, therefore, we get for the $x$ part of the wavefunction

$$X = \sqrt{\frac{2}{a}} \sin\left(\frac{n_x \pi x}{a}\right) \tag{3-23}$$

We can determine the form of the functions $Y$ and $Z$ in Eq. (3-11) in the same way; so the wavefunction for the system will be

$$\Psi = X(x)Y(y)Z(z)$$
$$= \sqrt{\frac{8}{V}} \sin\left(\frac{n_x \pi x}{a}\right) \sin\left(\frac{n_y \pi y}{b}\right) \sin\left(\frac{n_z \pi z}{c}\right) \tag{3-24}$$

where $V$ is the volume of the box ($V = abc$).
   The total energy $E$ is given by Eq. (3-15); therefore,

$$E = \frac{\pi^2 \hbar^2}{2m}\left(\frac{n_x^2}{a^2} + \frac{n_y^2}{b^2} + \frac{n_z^2}{c^2}\right) \tag{3-25}$$

## 3-3   EXAMINATION OF THE SOLUTION OF THE PARTICLE-IN-A-BOX PROBLEM

There are a couple of important observations we can immediately make regarding the solution of the particle-in-a-box problem:
   1. $n_x$, $n_y$, and $n_z$ can take any integral values. If the dimensions of the box are equal, i.e., $a = b = c$, then the energy we get when, for example, $n_x = 1$, $n_y = 1$, and $n_z = 2$ is the same as that obtained when $n_x = 1, n_y = 2$, and $n_z = 1$, or when $n_x = 2$, $n_y = 1$, and $n_z = 1$. We see this by substituting these values into Eq. (3-25). When the energies corresponding to different wavefunctions are the same we say that the functions are *degenerate*. In our example the degeneracy is three-fold because there are three functions corresponding to the same energy. In problems of chemical interest degeneracy crops up very frequently, so if you are not already familiar with this concept you should make careful note of it now.
   2. Let us examine what happens to our particle if it is allowed to move over all space, i.e., becomes free.

Look at the equation for the wavefunctions, Eq. (3-24); it is

$$\Psi = \sqrt{\frac{8}{V}} \sin\left(\frac{n_x \pi x}{a}\right) \sin\left(\frac{n_y \pi y}{b}\right) \sin\left(\frac{n_z \pi z}{c}\right)$$

We can give the particle its freedom by allowing the dimensions of the box — $a$, $b$, and $c$ — to approach infinity. This means that $V$, the volume of the box, tends to infinity. Consequently, $\Psi$ approaches zero. Therefore, $\Psi^2$, the probability function, also tends to zero. What does this mean? Is this reasonable? Yes, it is reasonable because it means that the probability of finding the particle in any small finite region in space tends to zero as we increase the space in which it can move: this is what we would expect. We do, however, run into some difficulties with normalization of the wavefunction for a free particle (cf. Exercise 3-4).

Now look at the energy of the free particle. The pertinent equation here is Eq. (3-25):

$$E = \frac{\pi^2 \hbar^2}{2m}\left(\frac{n_x^2}{a^2} + \frac{n_y^2}{b^2} + \frac{n_z^2}{c^2}\right)$$

As $a$, $b$, and $c$ approach infinity the energy levels get closer together; in the limit they merge into a continuum. What this means is that the energy of a free particle is not quantized. These results are in accord with spectroscopic evidence. For example, we know that an atom produces a line spectrum that eventually merges into a continuum. The continuum is due to the electron leaving the field of the atom and becoming free; its energy is then no longer quantized. Strictly speaking, of course, its energy is still quantized to some extent since it is always under the influence of every particle in the universe. However, outside the atom the restrictions on it are insignificant — this is why we observe what appears to be a continuum in the spectrum.

3. What happens when we deal with particles with large masses — for example, macroscopic particles like golf balls and planets? Look at Eq. (3-25) again and let $m$ become very large. The energy difference between successive quantum levels becomes exceedingly small. In the limit of infinite mass the energy levels form a continuum, i.e., the energy is unquantized in this case. In fact, for any macroscopic-sized particle the allowed energy levels are close enough together for us to consider that they form, for all practical purposes, a continuum — quantum effects are therefore quite insignificant for macroscopic-sized bodies.

## SUMMARY

1. In this chapter we applied quantum theory to a very simple problem — the particle-in-a-box problem.

2. We set up the Schroedinger equation for the problem.

3. We solved the equation, a partial differential equation, by a well-known method — *separation of variables*.

4. The important result to emerge was the equation for the energy of the particle:

$$E = \frac{\pi^2 \hbar^2}{2m}\left(\frac{n_x^2}{a^2} + \frac{n_y^2}{b^2} + \frac{n_z^2}{c^2}\right) \tag{3-25}$$

$n_x$, $n_y$, and $n_z$ are quantum numbers.

5. We saw that *quantization of energy, quantum numbers* and *zero-point energy* arise naturally from the mathematics of quantum theory.

6. We examined the case of a free particle by letting the dimensions of the box increase to infinity. We saw that the energy of a free particle is unquantized.

7. We also considered what happens when the mass of the particle approaches macroscopic proportions. Again, for all practical purposes, the energy is unquantized. Results 6 and 7 are in accord with what we can experimentally observe.

## EXERCISES

**3-1**  An electron is confined to a one-dimensional box 100 Å long. How many allowed energy levels are there for this electron between 10 eV and 12 eV?

**3-2**  An electron is confined to a one-dimensional path of 10 Å length. Where in its path is the electron most likely to be if it is in the state for which $n_x = 5$?

**3-3**  The butadiene molecule, illustrated in Fig. 3-3, is an example of a chemical system that, for some purposes, can be approximated by the particle-in-a-box model. The four $\pi$-electrons that are involved in the bonding between the p-orbitals of the carbon atoms can be considered to be free to move over the whole

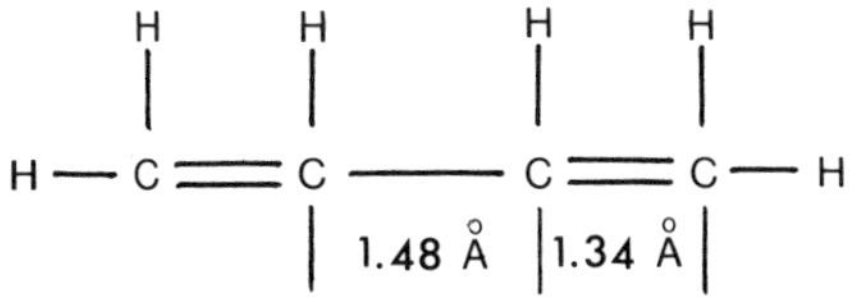

Fig. 3-3

length of the molecule. Though this is obviously a three-dimensional problem, it can be approximated by considering the $\pi$-electrons to be moving in a one-dimensional box whose length is the length of the molecule. Make the further approximation that the molecule is linear (it actually exists in cis and trans forms) and calculate the wavelength of the absorbed radiation when one of the $\pi$-electrons undergoes a transition from the ground state of the molecule to the first excited state. (Hint: Only two electrons can be accommodated in each energy level.)

**3-4**   When discussing the free particle in terms of the particle-in-a-box model, we mentioned that there would be difficulties with normalization of its wavefunction. Why is this so?

# 4

# General Formulation of the
# Theory of Quantum Mechanics

## 4-1 INTRODUCTION

We have previously derived the basic equation of quantum chemistry, the Schroedinger equation, from the equations of wave mechanics and the de Broglie relationship. We have shown that for a very simple system, such as a particle-in-a-box, this equation can be solved exactly. For more complex systems we cannot usually solve the Schroedinger equation exactly, so we have to resort to approximation methods. Before we outline these methods we shall, in effect, start our study of quantum mechanics again, deriving the Schroedinger equation from a set of seemingly arbitrary postulates. These postulates, however, are not as arbitrary as they at first sight seem; they are justified by the results they give. By this we mean that they explain the results of our experiments and observations on physical systems.

## 4-2 POSTULATES OF QUANTUM MECHANICS

There are several ways of choosing a set of postulates on which to base our quantum mechanical theory. Here we choose, for a start, the five listed below. Some authors of textbooks on quantum mechanics choose others, or combine the ones we have here to get fewer postulates — the final result is the same, and it can be shown that the various possible sets of postulates are equivalent. The postulates we list here are a convenient set for our present purposes; later we shall have to introduce a further postulate — the Pauli principle.

We consider a system of particles. This could be described classically at a time $t$ by giving values to the coordinates $q_k$ and the momenta $p_k$.

***Postulate 1***    The system can be described completely by a mathematical function $\Psi(q, t)$. $\Psi(q, t)$ is the wavefunction for the system and is interpreted in terms of $\Psi^*\Psi d\tau$, the probability of finding the value of $q_1$ to be between $q_1$ and $q_1 + dq_1$, $q_2$ to be between $q_2$ and $q_2 + dq_2$, $q_3$ to be between $q_3$ and $q_3 + dq_3$, etc., at a specific time $t$.

$\Psi(q, t)$ is a function of all the spatial coordinates and the time coordinate for the system. We note that $\Psi$ is a mathematical construct, useful for describing the system; i.e., what we have done is said that the state of the system can be expressed in mathematical language by the wavefunction $\Psi$. We can now use the rules of mathematical logic to obtain information about the system. However, we are not at liberty to give $\Psi$ any form. To be in accord with our physical interpretation in terms of probability (cf. Section 2-4) it must be everywhere finite, single-valued and square integrable. If these conditions are satisfied, then, as noted in the discussion of Section 2-4, we can measure the probability of finding the particles of our system with coordinates between $\tau$ and $\tau + d\tau$ by $\Psi^*\Psi d\tau$. (Note that we have written the coordinates in general form: $d\tau$ is an infinitesimal element of coordinate space; e.g., it stands for $dx_1 dy_1$-$dz_1 dx_2 dy_2 dz_2 \ldots dx_n dy_n dz_n$ for $n$ particles in a cartesian coordinate system.)

Before we leave this postulate we should examine the coordinate system needed to specify the state of our system. It is found that the number of variables needed for a complete description is equal to the number of classical degrees of freedom — only in this case can we describe macrobodies by both quantum mechanics and classical mechanics. Therefore, the number of spatial coordinates needed is $3n$ where $n$ is the number of particles in the system.

***Postulate 2***    Every observable physical quantity in the system has a linear Hermitian operator associated with it. To determine the operator corresponding to an observable we first write the classical representation of the observable in terms of the coordinates $q_k$ and the momenta $p_k$; we leave the coordinates alone and replace the momenta $p_k$ by $(\hbar/i)$ $(\partial/\partial q_k)$.

This postulate states that every observable physical quantity of a system can be represented by a mathematical operator. A mathematical operator is, as its name implies, something that operates on a mathematical function — it leaves it unchanged or changes it to another function.

Here are some examples of operators with the results of their effects on some arbitrarily chosen functions (the operators are in parentheses):

$$(3\cdot)x^4 = 3x^4$$
$$(x)x^5 = x^6$$
$$\left(\frac{d}{dx}\right)3x^2 = 6x$$

An important operator in quantum mechanics is the *Laplacian operator* which is usually given the symbol $\nabla^2$:

$$(\nabla^2)x^2y^2z^2 = \left(\frac{\partial^2}{\partial x^2}+\frac{\partial^2}{\partial y^2}+\frac{\partial^2}{\partial z^2}\right)x^2y^2z^2$$
$$= 2y^2z^2 + 2x^2z^2 + 2x^2y^2$$

In general we can say that if $\hat{A}$ is an operator (in this text we shall distinguish operators by placing circumflexes over them) then

$$\hat{A}f = g$$

$f$ and $g$ are mathematical functions.

By now you should see what we are getting at: we are going to operate on the wavefunction that describes our system with operators characterizing the observables (to save our constantly writing observable physical quantity, in future we shall abbreviate this to observable).

If we return to the postulate, we see that, in fact, we stated that the operators representing our observables should be linear operators. (We also stated that they should be Hermitian, but we shall leave the discussion of Hermiticity until we deal with the consequences of the postulates.) A *linear operator* has the properties

$$\hat{A}(f_1+f_2) = \hat{A}f_1 + \hat{A}f_2$$
$$\hat{A}kf = k\hat{A}f$$

Here $\hat{A}$ is a linear operator; $f, f_1$, and $f_2$ are functions, and $k$ is a constant.
An example of a linear operator is the differential operator; e.g.,

$$\frac{d}{dx}[f(x)+g(x)] = \frac{d}{dx}f(x) + \frac{d}{dx}g(x)$$

You will have gathered that not all operators are linear. A non-linear example is the square root operator:

$$\sqrt{f_1+f_2} \neq \sqrt{f_1} + \sqrt{f_2}$$

How do we decide the form of an operator representing an observable? There are rules for doing this, which are justified by the success achieved when we use them. Here are the rules for constructing operators: (Note that we have included these as part of our second postulate.)

1. If $q_k$ is a coordinate, e.g., $x$, $y$, or $z$, classically we represent this simply as $q_k$. In quantum mechanics we do the same thing, i.e., we leave the coordinates unchanged. We can write

$$\hat{q}_k \equiv q_k \tag{4-1}$$

2. If $p_k$ is a linear momentum, e.g., $p_x$, $p_y$ or $p_z$, the classical representation is

$$p_k = m\dot{q}_k$$

(We are using the conventional, short-hand form for derivatives with respect to time, i.e., $\dot{q} \equiv dq/dt$, $\ddot{q} \equiv d^2q/dt^2$, etc.) In quantum mechanics we replace this momentum by $(\hbar/i)(\partial/\partial q_k)$.

We can summarize these rules by saying that if we wish to determine the form of a quantum mechanical operator corresponding to an observable:

1. We find the classical representation of the observable in terms of the coordinates $q_k$ and the linear momenta $p_k$. (We can do this for all observables that appear in classical theory.)

2. We leave the coordinates alone.

3. We replace the momenta $p_k$ by $(\hbar/i)(\partial/\partial q_k)$.

As an example of this, consider a very important observable, angular momentum **M**. The component of angular momentum about the $x$-axis is given by

$$M_x = yp_z - zp_y$$

The quantum mechanical operator corresponding to this is

$$\hat{M}_x = \frac{\hbar}{i}\left(y\frac{\partial}{\partial z} - z\frac{\partial}{\partial y}\right) \tag{4-2}$$

***Postulate 3***    The wavefunction $\Psi(q, t)$ satisfies an equation of the form

$$\hat{H}\Psi(q, t) = -\frac{\hbar}{i}\frac{\partial}{\partial t}\Psi(q, t) \tag{4-3}$$

where $\hat{H}$ is the operator associated with the energy of the system.

We call this equation *Schroedinger's time-dependent equation*. $\Psi(q, t)$ is a function that is dependent on the time. The equation allows us to predict any future or past state of the system, provided we know the present state. Most of the material in this book does not require knowledge of this equation; Schroedinger's equation for a stationary state is the equation we shall nearly always need. We can obtain this latter equation from the more general time-dependent equation, but we shall not do this now; instead, we shall wait until we examine the consequences of the postulates, at which time we shall also examine the operator $\hat{H}$ in more detail. Here we just note that $\hat{H}$ is called the *Hamiltonian operator*.

**Postulate 4**   When we measure the values $a_\lambda$ of an observable corresponding to an operator $\hat{A}$ the results we obtain must satisfy the equation

$$\hat{A}\Psi_\lambda = a_\lambda\Psi_\lambda \tag{4-4}$$

where $\Psi_\lambda$ is the wavefunction for the state for which we measure the value $a_\lambda$.

Equation (4-4) says that the operator $\hat{A}$ acts on the function $\Psi_\lambda$ to give back the function multiplied by a constant, $a_\lambda$. In mathematics, equations of this type are well-known and are associated with *eigenvalue* problems. We say that the $\Psi_\lambda$ are *eigenfunctions* of the operator $\hat{A}$; $a_\lambda$ are the corresponding eigenvalues. The states of the system that are represented by the eigenfunctions are called *eigenstates*. The postulate says that if we make a single measurement of an observable associated with the operator $\hat{A}$, the only possible values we can obtain are the eigenvalues $a_\lambda$. Put another way: If we measure the observable $a$, we can get a precise value for it only if the system on which the measurement is being made is in an eigenstate of the operator corresponding to $a$. Later in this chapter (Section 4-9) we discuss the measurement of the average value of an observable when we make a series of measurements on a system. In the meantime, we note that we find the eigenvalues by solving the equation. The meaning of the equation is apparent when we know that the word eigenfunction comes from the German for "proper function"; it is a proper function because it satisfies an equation together with certain boundary conditions.

Let us give an example of an equation involving eigenfunctions and eigenvalues—by so doing we can illustrate the peculiarities of equations of this type. Consider

$$\frac{d}{dx}e^{kx} = ke^{kx}$$

$k$ is a constant. Here the operator is $d/dx$; it has eigenfunctions $\exp(kx)$ with corresponding eigenvalues $k$. The main point we should note is that though this at first sight looks like an ordinary, algebraic equation, it is not: we cannot cancel the $\exp(kx)$ from both sides since we would then be left with the operator $d/dx$ on the left side with nothing to operate on — in operator algebra we are not allowed to do this.

**Postulate 5**    If $\Psi_1$ and $\Psi_2$ are eigenfunctions, corresponding to eigenvalues $a_1$ and $a_2$, of some operator $\hat{A}$ (i.e., represent states of the system), then the function $\Psi$, where $\Psi = c_1\Psi_1 + c_2\Psi_2$ and $c_1$ and $c_2$ are constants, represents a state of the system for which the probability of observing the value $a_1$ is $c_1^*c_1$ and the probability of observing the value $a_2$ is $c_2^*c_2$.

We obtain the wavefunctions for a system by solving a linear differential equation, the eigenvalue equation, for the system. Now you will recall from your mathematics courses that if $\Psi_1$ and $\Psi_2$ are solutions of a linear differential equation, then other solutions can be obtained by taking linear combinations of $\Psi_1$ and $\Psi_2$. In our case this means there are further solutions

$$\Psi = c_1\Psi_1 + c_2\Psi_2 \tag{4-5}$$

where $c_1$ and $c_2$ are constants. (This is known as the *principle of superposition*.) What we are saying here is that $\Psi$ represents a state of the system in which there is a certain probability of observing the value $a_1$ for the observable $a$ and a certain probability of observing the value $a_2$ — these probabilities are related to the coefficients in the linear combination.

Now though $\Psi$ represents a state of the system, this state is not in general an eigenstate — this we can very easily prove. Operating on both sides of Eq. (4-5) with $\hat{A}$, we get

$$\hat{A}\Psi = \hat{A}(c_1\Psi_1 + c_2\Psi_2) \tag{4-6}$$

But

$$\hat{A}\Psi_1 = a_1\Psi_1 \quad \text{and} \quad \hat{A}\Psi_2 = a_2\Psi_2$$

Therefore,

$$\hat{A}\Psi = c_1 a_1 \Psi_1 + c_2 a_2 \Psi_2 \tag{4-7}$$

This shows that we cannot in general write

$$\hat{A}(c_1\Psi_1 + c_2\Psi_2) = a(c_1\Psi_1 + c_2\Psi_2) \tag{4-8}$$

where $a$ is a constant; i.e., $\Psi$ is not in general an eigenfunction of $\hat{A}$.

If $a_1 = a_2$, we *could* write an equation of the form Eq. (4-8). In this case, when the eigenvalues corresponding to different eigenfunctions are the same, we say that the functions are degenerate (cf. Section 3-3). We can therefore conclude that linear combinations of degenerate eigenfunctions of an operator $\hat{A}$ are also eigenfunctions of $\hat{A}$. This is an important result.

Referring back to Eq. (4-5) and anticipating a result we are going to obtain in Section 4-5, we can easily show that

$$c_1^* c_1 + c_2^* c_2 = 1 \tag{4-9}$$

The result we are going to anticipate is a very important property of eigenfunctions of quantum mechanical operators. The property is *orthogonality*. Very briefly, this means that if $\Psi_1$ and $\Psi_2$ are eigenfunctions of a quantum mechanical operator with different eigenvalues, then

$$\int \Psi_1^* \Psi_2 d\tau = \int \Psi_2^* \Psi_1 d\tau = 0 \tag{4-10}$$

We mention this property here since it is needed to prove Eq. (4-9). In Section 4-5 we discuss orthogonality of eigenfunctions in more detail. In the meantime, we shall continue with the proof of Eq. (4-9).

For $\Psi$ to be normalized we must have

$$\int \Psi^* \Psi d\tau = 1$$

Therefore, since $\Psi$ is given by Eq. (4-5), we get

$$\int (c_1^* \Psi_1^* + c_2^* \Psi_2^*)(c_1 \Psi_1 + c_2 \Psi_2) d\tau = c_1^* c_1 \int \Psi_1^* \Psi_1 d\tau$$

$$+ c_1^* c_2 \int \Psi_1^* \Psi_2 d\tau + c_2^* c_1 \int \Psi_2^* \Psi_1 d\tau + c_2^* c_2 \int \Psi_2^* \Psi_2 d\tau = 1 \tag{4-11}$$

(We have taken $c_1$, $c_2$, $c_1^*$, and $c_2^*$ outside the integrals because they are constants.) Now $\Psi_1$ and $\Psi_2$ are normalized functions, i.e.,

$$\int \Psi_1^* \Psi_1 d\tau = \int \Psi_2^* \Psi_2 d\tau = 1$$

Substituting this and Eq. (4-10) into Eq. (4-11), we get

$$c_1^* c_1 + c_2^* c_2 = 1$$

This is what we set out to prove.

This completes our initial examination of this postulate. There is a very important consequence concerning measurement of average values of observables—we shall examine this as a general consequence of the postulates.

## General Consequences of the Postulates

Here we shall examine some immediate consequences of the postulates we have made. These consequences follow either from the fact that we are dealing with physically real systems, or by mathematical logic. The point we are making is that we do not need to make any further postulates for the moment.

## 4-3  NORMALIZATION OF THE WAVEFUNCTION

This we have mentioned several times already. Here we shall very briefly repeat what we have already said about normalization in order to keep our general approach to quantum mechanics complete.

The mathematical expression of normalization is

$$\int \Psi^*\Psi d\tau = 1 \qquad (4\text{-}12)$$

In practice it is often convenient to use unnormalized wavefunctions, normalizing them only when we wish to interpret our results in terms of our physical system; so you should not be disturbed if in later chapters we work with unnormalized functions.

## 4-4  HERMITICITY OF OPERATORS

We now discuss a very important property of quantum mechanical operators — *Hermiticity*. All the operators we use in quantum mechanics belong to a class of operators that are defined as being *Hermitian*. At this level of discussion of quantum mechanics we do not need to go into the details of the meaning of Hermiticity; it is sufficient for us to accept the mathematical definition, which is: An operator $\hat{A}$ is said to be Hermitian if it satisfies the following relationship for the functions $\Psi_1$ and $\Psi_2$:

$$\int \Psi_1^*\hat{A}\Psi_2 d\tau = \int \Psi_2 \hat{A}^*\Psi_1^* d\tau \qquad (4\text{-}13)$$

Here the functions $\Psi_1$ and $\Psi_2$ are acceptable functions (they are acceptable in the sense that they can represent physical states of the system, i.e., they are finite, single-valued and are square integrable). $\Psi_1$ and $\Psi_2$ are defined over a certain range of coordinate space, and the integrals are evaluated over this range. $\hat{A}$ is a Hermitian operator.

The significance of this is that eigenvalues of Hermitian operators are real, a statement we prove in a moment. Since observables must be real, it is necessary that the operators associated with them give real eigenvalues — so long as these operators are Hermitian, we are assured of real eigenvalues. We can, in fact, say that Hermiticity is to operators what reality is to numbers.

The proof that eigenvalues of Hermitian operators are always real is quite simple. We shall work through it: The eigenvalues of an operator $\hat{A}$ are defined by Eq. (4-4):

$$\hat{A}\Psi_\lambda = a_\lambda\Psi_\lambda$$

Taking the complex conjugate of this equation, we get

$$\hat{A}^*\Psi_\lambda^* = a_\lambda^*\Psi_\lambda^* \tag{4-14}$$

Multiply Eq. (4-4) from the left by $\Psi_\lambda^*$; multiply Eq. (4-14) from the left by $\Psi_\lambda$; we get

$$\Psi_\lambda^*\hat{A}\Psi_\lambda = \Psi_\lambda^* a_\lambda\Psi_\lambda \tag{4-15}$$

$$\Psi_\lambda\hat{A}^*\Psi_\lambda^* = \Psi_\lambda a_\lambda^*\Psi_\lambda^* \tag{4-16}$$

Now integrate these two equations over the volume of coordinate space:

$$\int \Psi_\lambda^*\hat{A}\Psi_\lambda d\tau = a_\lambda \int \Psi_\lambda^*\Psi_\lambda d\tau \tag{4-17}$$

$$\int \Psi_\lambda\hat{A}^*\Psi_\lambda^* d\tau = a_\lambda^* \int \Psi_\lambda\Psi_\lambda^* d\tau \tag{4-18}$$

Here we have been able to take $a_\lambda$ and $a_\lambda^*$ outside the integral signs because they are numbers, not operators. We cannot, however, take the operators $\hat{A}$ and $\hat{A}^*$ from the integrals.

The left sides of Eqs. (4-17) and (4-18) are equal – due to the Hermitian property of $\hat{A}$ (cf. Eq. (4-13)); therefore

$$a_\lambda = a_\lambda^*$$

i.e., all the eigenvalues $a_\lambda$ are real.

One point we should make before we leave this discussion is this: If we obtain a quantum mechanical operator by applying our rules, i.e., taking the classical expression and substituting into it, we must be careful to arrange the terms so that the resulting operator is Hermitian. For example, direct application of the rules for forming operators to the expression $xp_x$ gives $(\hbar/i) x(\partial/\partial x)$ – this is not Hermitian. However, if we write the classical expression as $\frac{1}{2}(xp_x + p_x x)$, then apply the rules, the resulting operator is Hermitian.

## 4-5  ORTHOGONALITY OF WAVEFUNCTIONS

Here we consider a very important property of different eigenfunctions of Hermitian operators, in particular those operators we use in quantum mechanics – orthogonality.

Let $\Psi_1$ and $\Psi_2$ be eigenfunctions of a Hermitian operator $\hat{A}$ with different eigenvalues $a_1$ and $a_2$; i.e., the equation

$$\hat{A}\Psi_\lambda = a_\lambda \Psi_\lambda \tag{4-4}$$

is satisfied by $\lambda = 1$ and $\lambda = 2$. We can now show that

$$\int \Psi_1^* \Psi_2 d\tau = 0 \tag{4-19}$$

We say that the eigenfunctions $\Psi_1$ and $\Psi_2$ are *orthogonal*. To prove this we substitute the eigenfunctions into Eq. (4-4); we get

$$\hat{A}\Psi_1 = a_1 \Psi_1 \tag{4-20}$$

$$\hat{A}\Psi_2 = a_2 \Psi_2 \tag{4-21}$$

Take the complex conjugate of Eq. (4-20):

$$\hat{A}^* \Psi_1^* = a_1 \Psi_1^* \tag{4-22}$$

We have replaced $a_1^*$ by $a_1$ since $a_1$ is a real eigenvalue, i.e., $a_1^* = a_1$. Multiply Eq. (4-22) from the left by $\Psi_2$; multiply Eq. (4-21) from the left by $\Psi_1^*$; we get

$$\Psi_2 \hat{A}^* \Psi_1^* = \Psi_2 a_1 \Psi_1^* \tag{4-23}$$

$$\Psi_1^* \hat{A} \Psi_2 = \Psi_1^* a_2 \Psi_2 \tag{4-24}$$

Now integrate both equations over all space and subtract one from the other:

$$\int \Psi_2 \hat{A}^* \Psi_1^* d\tau - \int \Psi_1^* \hat{A} \Psi_2 d\tau = (a_1 - a_2) \int \Psi_2 \Psi_1^* d\tau \tag{4-25}$$

Because $\hat{A}$ is a Hermitian operator, the left side of Eq. (4-25) is zero; therefore

$$(a_1 - a_2) \int \Psi_2 \Psi_1^* d\tau = 0 \tag{4-26}$$

Recalling that we designated $a_1$ and $a_2$ to be different eigenvalues, i.e., $a_1 \neq a_2$, we can now write

$$\int \Psi_2 \Psi_1^* d\tau = 0$$

This verifies our statement about orthogonality of wavefunctions.

We should note that if we had considered a degenerate state of the system, the wavefunctions corresponding to this state would not necessarily have to be orthogonal. In the case of degeneracy $a_1 = a_2$; the right side of Eq. (4-25) does not require the wavefunctions to be orthogonal to be equal to zero.

What does the orthogonality condition mean? We can understand this by drawing an analogy with vectors. In mathematics we define the *scalar product* of two vectors as

$$\mathbf{a \cdot b} = ab \cos \theta$$

$a$ and $b$ are the magnitudes of the vectors; $\theta$ is the angle between them. If this angle is 90°, then $\mathbf{a \cdot b} = 0$. The vectors are said to be *orthogonal*. What this means is that they are completely independent of each other. Going back to our wavefunctions, we can say that here too orthogonality implies complete independence of each other.

## 4-6  DIRAC NOTATION

Whilst we are making an analogy between wavefunctions and vectors we can mention that it is possible to represent the state of a system by a *state vector*. The notation associated with this representation is due to Dirac. We shall not use Dirac notation in this book; however, it is commonly used in modern texts on quantum chemistry and in quantum chemical papers, so it would be well for you to know how it corresponds to our notation.

In our formulation of quantum mechanics we have been using $\Psi(x)$, for example, to denote a wavefunction. In Dirac notation we write instead of this the following: $\langle x|\Psi \rangle$. If we do not wish to show the $x$ dependence, we can abbreviate this to $|\Psi \rangle$ — this is called a *ket vector*. Corresponding to the complex conjugate wavefunction, we write $\langle \Psi|x \rangle$, or, more simply, just $\langle \Psi|$. This latter vector therefore corresponds to $\Psi^*(x)$ — it is called a *bra vector*.

Let us now very briefly examine how Dirac notation would be used to express some of the theory we have so far formulated:

1. The normalization condition is written

$$\int \langle \Psi|x \rangle \langle x|\Psi \rangle \, dx = 1$$

We can abbreviate this to

$$\langle \Psi|\Psi \rangle = 1$$

2. A very commonly occurring integral in our notation is

$$\int \Psi_1^* \hat{A} \Psi_2 dx$$

In the abbreviated notation this becomes

$$\langle \Psi_1|\hat{A}|\Psi_2 \rangle$$

3. Finally, we shall express the Hermiticity condition; it is

$$\langle \Psi_1 | \hat{A} | \Psi_2 \rangle = \langle \Psi_2 | \hat{A} | \Psi_1 \rangle *$$

What we have said here should be sufficient to indicate how to translate Dirac notation into our notation. This is about all we need to know about the former notation at this stage.

## 4-7  ORTHONORMALITY AND COMPLETENESS OF WAVEFUNCTIONS

When functions are both orthogonal and normalized we say that they are *orthonormal*. We can express this conveniently by using a well-known mathematical symbol, the *Kronecker delta* $\delta_{ij}$. Using this, we can write

$$\int \Psi_i^* \Psi_j d\tau = \delta_{ij} \tag{4-27}$$

The Kronecker delta equals zero unless $i = j$; in this latter case $\delta_{ii} = 1$. We see then that Eq. (4-27) merely expresses what we have previously used two equations for (Eqs. (4-12) and (4-19)).

If we have a set of functions that are all orthogonal to each other and normalized, we say that we have an *orthonormal set*. Furthermore, if there is no other function with the same boundary conditions that is orthogonal to any member of the set, we call the set a *complete orthonormal set*.

So far as we are concerned, the meaning of a complete set of orthonormal functions is that none of them contains any component of the others; also, it is not possible to find any other orthogonal functions in the system. We can see this intuitively by using the analogy with vectors again. An orthonormal set of functions is analogous to a set of orthogonal vectors, each vector being of unit length. In three-dimensional space, for example, there are three, and only three, vectors that are mutually perpendicular. We can, admittedly, orient these vectors in any way we like, but we cannot find more than three. Extending this concept of orthogonality into higher-dimensioned spaces, we see that there are $n$, and only $n$, orthogonal vectors in an $n$-dimensional space. We cannot picture this extension of the concept because we live in a three-dimensional world, but this does not prevent our using it in mathematical problems.

Now, although we have seen that orthogonal functions are completely independent of each other by analogizing with vectors, we can, in fact, prove this mathematically. This we now do:

Let us try to write one eigenfunction $\Psi_1$ of the complete orthonormal set as a linear combination of all the other members of the set:

$$\Psi_1 = c_2\Psi_2 + c_3\Psi_3 + \cdots + c_n\Psi_n = \sum_{i=2}^{n} c_i\Psi_i \qquad (4\text{-}28)$$

$c_2$, $c_3$, etc., are constants. Multiply by $\Psi_j^*$, where $j$ can take all integral values except 1, i.e., $j = 2, 3, \ldots, n$; we get

$$\Psi_j^*\Psi_1 = c_2\Psi_j^*\Psi_2 + c_3\Psi_j^*\Psi_3 + \cdots + c_n\Psi_j^*\Psi_n \qquad (4\text{-}29)$$

Integrate both sides of this equation over all coordinate space:

$$\int \Psi_j^*\Psi_1 d\tau = c_2 \int \Psi_j^*\Psi_2 d\tau + c_3 \int \Psi_j^*\Psi_3 d\tau + \cdots + c_n \int \Psi_j^*\Psi_n d\tau \qquad (4\text{-}30)$$

Because the wavefunctions are orthogonal, the left side of Eq. (4-30) is zero for all values of $j$ (remember that $j \neq 1$). Looking at the right side, we see that the only integral that is not zero is the one where $j$ has the value of the subscript of the second function under the integral — this is also because of the orthogonality of the functions. Since the functions are normalized, the non-zero integral equals one. (This latter statement is, of course, incidental to this proof — it would not matter if the functions were not normalized.) Therefore,

$$0 = c_j \cdot 1$$

i.e.,

$$c_j = 0$$

for all values of $j$. This proves what we said: $\Psi_1$ contains no components of the other members of the orthonormal set to which it belongs, i.e., it is completely independent of these functions.

A question we should ask here is: Do eigenfunctions of the operators we use in quantum chemistry form complete orthonormal sets? The answer, if we ignore the possibility of degeneracy, is yes. (And even in the case of degeneracy we can always find a set of orthogonal functions by taking linear combinations of the degenerate functions such that the linear combinations are orthogonal to each other.) All the operators we use in quantum chemistry are Hermitian. We have already seen (Section 4-5) that the eigenfunctions of a Hermitian operator are orthogonal, and we can always normalize these functions to get an orthonormal set. So we are just left with the completeness condition to consider. Actually, we cannot give a rigorous proof of the completeness condition. However, it is true for all the operators of interest in quantum chemistry, so we shall assume this condition when it is needed — no difficulties arise because of this assumption.

## 4-8  EXPANSION OF WAVEFUNCTIONS

Here we shall show the usefulness of the completeness condition we have just been discussing. We shall show the basis for a well-used technique in solving quantum mechanical problems — expansion of an arbitrary function in terms of the members of a complete orthonormal set of functions.

Let the arbitrary function be $\Psi$. Let the members of the complete set of orthonormal functions with the same boundary conditions as $\Psi$ be $\Psi_1, \Psi_2, \ldots, \Psi_n$. We can then write $\Psi$ as a linear combination of these functions:

$$\Psi = c_1\Psi_1 + c_2\Psi_2 + \cdots + c_n\Psi_n \tag{4-31}$$

$c_1, c_2, \ldots, c_n$ are constants. This is equivalent to writing a vector in terms of its components along the coordinate axes. For example, consider the diagram in Fig. 4-1. We write the vector **V** from the origin to the point $P$ in

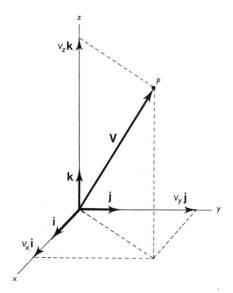

**Fig. 4-1**  Components of a vector **V** along the coordinate axes.

three-dimensional space as

$$\mathbf{V} = V_x\mathbf{i} + V_y\mathbf{j} + V_z\mathbf{k}$$

$\mathbf{i}, \mathbf{j},$ and $\mathbf{k}$ are unit vectors along the coordinate axes; $V_x, V_y,$ and $V_z$ are the components of the vector in the directions of the axes. We can extend this treatment to other-dimensioned spaces. All we are doing in Eq. (4-31), then, is continuing our analogy of wavefunctions with vectors.

How do we find the values of the coefficients $c_1, c_2, \ldots, c_n$? This is easy: Multiply both sides of Eq. (4-31) by $\Psi_j^*$ ($j$ takes all values 1 to $n$); we get

$$\Psi_j^*\Psi = c_1\Psi_j^*\Psi_1 + c_2\Psi_j^*\Psi_2 + \cdots + c_n\Psi_j^*\Psi_n \qquad (4\text{-}32)$$

Now integrate over all coordinate space:

$$\int \Psi_j^*\Psi \, d\tau = c_1 \int \Psi_j^*\Psi_1 d\tau + c_2 \int \Psi_j^*\Psi_2 d\tau + \cdots + c_n \int \Psi_j^*\Psi_n d\tau \qquad (4\text{-}33)$$

The only term on the right side of Eq. (4-33) that we are left with is the one where $j$ takes the value of the subscript of the second function under the integral sign; all others are zero because of the orthogonality of the functions. Also, we see that our remaining integral is unity because of the normality condition. Therefore, we can write instead of Eq. (4-33)

$$\int \Psi_j^*\Psi \, d\tau = c_j \qquad (4\text{-}34)$$

In words, we now have an equation for finding the values of the coefficients in the expansion Eq. (4-31). In later chapters we shall see the great use of this method of expressing wavefunctions.

## 4-9   AVERAGE VALUES OF OBSERVABLES

We can use the expansion of Eq. (4-31) to get another important equation — the equation for the average value of an observable.

Let $\Psi$ be a wavefunction for a system; let $\Psi_1, \Psi_2, \ldots, \Psi_n$ be the complete orthonormal set of eigenfunctions for an operator $\hat{A}$ corresponding to some observable in the system. We can expand $\Psi$ in terms of the set, according to Eq. (4-31):

$$\Psi = c_1\Psi_1 + c_2\Psi_2 + \cdots + c_n\Psi_n \qquad (4\text{-}35)$$

$c_1, c_2,$ etc. are constants. Operate on both sides of Eq. (4-35) from the left with $\hat{A}$; we get

$$\hat{A}\Psi = c_1\hat{A}\Psi_1 + c_2\hat{A}\Psi_2 + \cdots + c_n\hat{A}\Psi_n \qquad (4\text{-}36)$$

(The order of $\hat{A}$ and the constants can be changed since $\hat{A}$ is linear.) But

$\Psi_1$, $\Psi_2$, etc. are eigenfunctions of $\hat{A}$. So we can write for Eq. (4-36)

$$\hat{A}\Psi = c_1 a_1 \Psi_1 + c_2 a_2 \Psi_2 + \cdots + c_n a_n \Psi_n \qquad (4\text{-}37)$$

where the $a$'s are the eigenvalues corresponding to the eigenfunctions. Now go back to Eq. (4-35) and take the complex conjugate of both sides; we get

$$\Psi^* = c_1^* \Psi_1^* + c_2^* \Psi_2^* + \cdots + c_n^* \Psi_n^* \qquad (4\text{-}38)$$

Multiply the left side of Eq. (4-37) by the left side of Eq. (4-38); multiply the right side of Eq. (4-37) by the right side of Eq. (4-38). After integrating over all coordinate space, we get

$$\int \Psi^* \hat{A} \Psi d\tau = c_1^* c_1 a_1 \int \Psi_1^* \Psi_1 d\tau + c_2^* c_2 a_2 \int \Psi_2^* \Psi_2 d\tau$$
$$+ \cdots + c_n^* c_n a_n \int \Psi_n^* \Psi_n d\tau \qquad (4\text{-}39)$$

In this equation we have got rid of all the terms of the type $c_i^* c_j a_j \int \Psi_i^* \Psi_j d\tau$; these are all zero because of the orthogonality of the eigenfunctions. Only when $i = j$ are the terms not zero—these are the ones we have retained. Now the integrals on the right side of Eq. (4-39) are each equal to one, because of the normality condition. Therefore, we write

$$\int \Psi^* \hat{A} \Psi d\tau = c_1^* c_1 a_1 + c_2^* c_2 a_2 + \cdots + c_n^* c_n a_n \qquad (4\text{-}40)$$

By postulate 5 the right side of Eq. (4-40) is the average value of the observable $a$ corresponding to the operator when the system is in the state $\Psi$. Using $\bar{a}$ for this average value, we can write the important expression

$$\bar{a} = \int \Psi^* \hat{A} \Psi d\tau \qquad (4\text{-}41)$$

Here we have defined the average value as in statistics: if we have a large number of measurements $a_1$, $a_2$, etc. of a physical quantity, then the average value is determined by summing the measurements and dividing by the total number of measurements.

Note that we have arrived at Eq. (4-41) by using postulate 5. Some authors make the expression for the average value of an observable the postulate: this indicates the importance of it. This expression allows us to translate back from our quantum mechanical theory, which is essentially mathematical, to our physically real systems; it tells us the result we should expect if we average the results of a large number of measurements of an observable.

## 4-10   COMMUTATION OF OPERATORS

We consider two operators $\hat{A}$ and $\hat{B}$. Allow these to operate successively on a function $\Psi$. Does the order in which they operate matter? i.e., is the result of $\hat{A}\hat{B}\Psi$ the same as $\hat{B}\hat{A}\Psi$? The answer to this depends on the nature of the operators. If, in fact, we do get the same result in both cases (i.e., $\hat{A}\hat{B}\Psi = \hat{B}\hat{A}\Psi$), we say that the operators *commute*. As an example of commuting operators we can consider the operators $5 \cdot$ and $d/dx$ acting on the function $\exp(x)$:

$$5 \cdot \frac{d}{dx}(e^x) = 5e^x$$

(When we see two or more operators in succession, the operations must be performed from right to left. Here this means $d/dx$ first, then $5 \cdot$.)

$$\frac{d}{dx} \cdot 5 \cdot (e^x) = 5e^x$$

i.e., the order in which we perform the operations does not matter, the operators $5 \cdot$ and $d/dx$ commute.

On the other hand we can illustrate non-commuting operators by considering $x \cdot$ and $d/dx$ acting successively on $\exp(x)$:

$$x \cdot \frac{d}{dx}(e^x) = xe^x$$

$$\frac{d}{dx}x \cdot (e^x) = \frac{d}{dx}(xe^x) = xe^x + e^x$$

Clearly, we do not get the same result: the operators $x \cdot$ and $d/dx$ do not commute.

To summarize, we can say that if two operators $\hat{A}$ and $\hat{B}$ commute, then

$$\hat{A}\hat{B} - \hat{B}\hat{A} = 0 \qquad\qquad (4\text{-}42)$$

You will often see this written in the abbreviated form

$$\{\hat{A}, \hat{B}\} = 0$$

The brackets { } are called *commutator brackets*. We shall not use this notation in this book, but shall write commutation relations in full for clarity.

## 4-11  QUANTUM MECHANICAL MEANING OF COMMUTATION OF OPERATORS

Commutation of operators has an extremely important meaning in quantum mechanics: if two operators representing observables commute, we can find a set of functions that are simultaneously eigenfunctions of both operators. We shall prove this, then examine its meaning.

Let the eigenfunctions of an operator $\hat{A}$ be $\theta_i$; let the eigenfunctions of an operator $\hat{B}$ be $\chi_j$. This means that we can write

$$\hat{A}\theta_i = a_i\theta_i \tag{4-43}$$

$$\hat{B}\chi_j = b_j\chi_j \tag{4-44}$$

$a_i$ and $b_j$ are eigenvalues corresponding to the eigenfunctions $\theta_i$ and $\chi_j$. Multiply Eq. (4-44) from the left by $\hat{A}$; we get

$$\hat{A}\hat{B}\chi_j = \hat{A}b_j\chi_j \tag{4-45}$$

But $b_j$ is just a constant; therefore

$$\hat{A}\hat{B}\chi_j = b_j\hat{A}\chi_j \tag{4-46}$$

Now remember that the operators commute. We can therefore write instead of Eq. (4-46)

$$\hat{B}(\hat{A}\chi_j) = b_j(\hat{A}\chi_j) \tag{4-47}$$

This is another eigenvalue equation. It says that $\hat{A}\chi_j$ is an eigenfunction of the operator $\hat{B}$ with eigenvalue $b_j$. Now we know that $\chi_j$ is an eigenfunction of $\hat{B}$ with eigenvalue $b_j$ (cf. Eq. (4-44)); this tells us that $\chi_j$ is also an eigenfunction of $\hat{A}$. Why can we say this? There are two possibilities:

1. If $\chi_j$ is a non-degenerate function, $\hat{A}\chi_j$ can only differ from $\chi_j$ by a constant multiplier, $k$; i.e.,

$$\hat{A}\chi_j = k\chi_j \tag{4-48}$$

Therefore, $\chi_j$ is an eigenfunction of $\hat{A}$: it must be one of the set $\theta_i$; i.e., there are simultaneous eigenstates for both operators.

2. If $\chi_j$ is a member of a degenerate set of eigenfunctions, then $\hat{A}\chi_j$ can be expressed as a linear combination of these functions:

$$\hat{A}\chi_j = k\chi_j + k_1\chi_{j1} + k_2\chi_{j2} + \cdots \tag{4-49}$$

$\chi, \chi_{j1}, \chi_{j2}$, etc. are the members of the degenerate set; $k, k_1, k_2$, etc. are constants. We can always find a linear combination, like that in Eq.

(4-49), that will be an eigenfunction of $\hat{A}$. Let us consider the doubly degenerate case as an example. (A similar argument applies when the degeneracy is greater than this.) We call the degenerate functions $\chi_j$ and $\chi_{j1}$. This allows us to write

$$\hat{A}\chi_j = a\chi_j + b\chi_{j1} \tag{4-50}$$

$$\hat{A}\chi_{j1} = c\chi_j + d\chi_{j1} \tag{4-51}$$

$a$, $b$, $c$, and $d$ are constants. We define some new constants $k$, $l$, $m$, and $n$ by means of the equations

$$a = \frac{mk - lc}{k} \qquad b = \frac{ml - ld}{k}$$

$$c = \frac{la - nl}{k} \qquad d = \frac{nk + lb}{k}$$

Substitute for $a$ and $b$ in Eq. (4-50); we get

$$\hat{A}k\chi_j = (mk - lc)\chi_j + (ml - ld)\chi_{j1} \tag{4-52}$$

Multiply Eq. (4-51) by $l$; we get

$$\hat{A}l\chi_{j1} = lc\chi_j + ld\chi_{j1} \tag{4-53}$$

Adding Eqs. (4-52) and (4-53) gives

$$\hat{A}(k\chi_j + l\chi_{j1}) = m(k\chi_j + l\chi_{j1}) \tag{4-54}$$

By substituting for $c$ and $d$ in Eq. (4-51), we get

$$\hat{A}(l\chi_j - k\chi_{j1}) = n(l\chi_j - k\chi_{j1}) \tag{4-55}$$

Equations (4-54) and (4-55) are eigenvalue equations; they define eigenfunctions of $\hat{A}$. Therefore, we have shown that there are linear combinations of the degenerate set of eigenfunctions of $\hat{B}$ that are eigenfunctions of $\hat{A}$: These combinations must belong to the set $\theta_i$. But linear combinations of degenerate eigenfunctions represent eigenstates of the system (cf. Section 4-2, postulate 5). Therefore, we have proved that there are simultaneous eigenstates for the degenerate case also.

The converse of what we have just been discussing is also true: If eigenstates exist simultaneously for two operators, then these operators must commute. This is very easy to prove:

Let $\hat{A}$ and $\hat{B}$ be the operators in question. If we are to know eigenstates for these simultaneously, there must be a function $\Psi$ that satisfies the equations for both $\hat{A}$ and $\hat{B}$; i.e.,

$$\hat{A}\Psi = a\Psi \qquad (4\text{-}56)$$

$$\hat{B}\Psi = b\Psi \qquad (4\text{-}57)$$

$a$ and $b$ are constants, the eigenvalues corresponding to the eigenfunction $\Psi$. Operate on Eq. (4-56) from the left with $\hat{B}$:

$$\hat{B}\hat{A}\Psi = \hat{B}a\Psi \qquad (4\text{-}58)$$

$a$ is just a constant, and $\hat{B}$ is a linear operator; therefore, we can write

$$\hat{B}a\Psi = a\hat{B}\Psi = ab\Psi \qquad (4\text{-}59)$$

Similarly, if we operate on Eq. (4-57) from the left with $\hat{A}$ we get

$$\hat{A}\hat{B}\Psi = \hat{A}b\Psi = b\hat{A}\Psi = ba\Psi = ab\Psi \qquad (4\text{-}60)$$

$a$ and $b$ are just constants, so we can change the order of multiplication. We have therefore proved that

$$\hat{B}\hat{A}\Psi = \hat{A}\hat{B}\Psi \qquad (4\text{-}61)$$

i.e., the operators must commute.

Before we leave this section we should mention that analogous theorems to those we have just proved apply when we consider more than two operators: simultaneous eigenstates exist for all operators that commute with each other, and if simultaneous eigenstates exist for a set of operators, then these operators must commute.

## 4-12 HEISENBERG UNCERTAINTY PRINCIPLE

What does the discussion of the last section mean in physical terms? The answer to this is of great consequence, for it points out one of the very fundamental differences between classical mechanics and quantum mechanics: it shows us how our postulates are in accord with the *Heisenberg uncertainty principle* [3].

The fact that eigenstates exist simultaneously for two observables means that we can measure both these observables precisely at any given moment in time. Simultaneous eigenstates are possible only if the operators for the observables in question commute; if the operators do not commute, then we cannot measure these observables precisely at the same time. As an example we can consider the possibility of measuring the position of a particle and its momentum. The operator for position is $x$ ·. (We shall consider only one dimension for the sake of this argument.) The operator for linear momentum $p_x$ is $(\hbar/i)(\partial/\partial x)$. These do not commute, as we can very easily show:

$$\frac{\hbar}{i}\frac{\partial}{\partial x}x \cdot = \frac{\hbar}{i}\left(x\frac{\partial}{\partial x}+1\cdot\right) = x\frac{\hbar}{i}\frac{\partial}{\partial x}+\frac{\hbar}{i}$$

Therefore,

$$\frac{\hbar}{i}\frac{\partial}{\partial x}x \cdot - x\frac{\hbar}{i}\frac{\partial}{\partial x} = \frac{\hbar}{i} \qquad (4\text{-}62)$$

In fact, from the relation Eq. (4-62) we could show, though we shall not do so, that

$$\Delta x\Delta p_x \geqslant \frac{\hbar}{2} \qquad (4\text{-}63)$$

$\Delta x$ is the uncertainty in position; $\Delta p_x$ is the uncertainty in the linear momentum in the $x$ direction (we are still talking about one dimension)[†].

Equation (4-63) is an expression of the uncertainty principle. Heisenberg put forward this principle in 1927 and thereby stated one of the major differences between classical and quantum mechanics. Equation (4-63) is a natural consequence of quantum theory. The question we should ask is: Is this uncertainty relationship in accord with what we can experimentally observe? After all, a theory is useful only if it is in accord with physical reality. To examine the question we must discuss the possibility of measuring the position and momentum of a particle (an electron, for example) at the same time.

Consider the experimental set-up in Fig. 4-2. The electron is moving in the positive $y$ direction toward the very small slit of width $\Delta x$. Before the electron reaches the slit the $x$ component of its momentum is zero, i.e., $p_x = 0$. Now since the electron has a wave nature, we might expect it to be diffracted at the slit, in the same way as light waves would be. The result of the diffraction is that the new direction of the momentum will not be precisely known; the maximum accuracy with which we can define the new direction is to say that it lies somewhere within the angles $\pm\theta$, where $\theta$ is given by the theory for diffraction of light waves[‡]:

$$\sin\theta = \frac{\lambda}{\Delta x} \qquad (4\text{-}64)$$

---

†For proof of Eq. (4-63) you can refer to reference E2 in the bibliography, p. 348.

‡See, for example, Miller, F. M. Jr., *College Physics*, Harcourt Brace and World, New York, 1959, pp. 451–452.

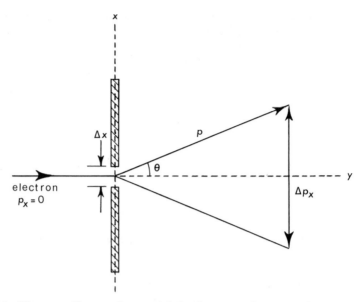

**Fig. 4-2**  Diagram to illustrate the uncertainties that occur when attempting to measure the position and momentum of an electron.

$\lambda$ is the wavelength associated with the electron. Whereas before diffraction the $x$ component of the momentum was zero, after diffraction there is an uncertainty in our knowledge of its value — this uncertainty is shown as $\Delta p_x$ in Fig. 4-2. It is evident from Fig. 4-2 that

$$\Delta p_x = 2p \sin \theta \qquad (4\text{-}65)$$

This equation defines the limits of our knowledge of the component of the electron's momentum. The limits of our knowledge of where the electron is on the $x$-axis at the slit are defined by the width of the slit, i.e., $\Delta x$. Therefore, using Eq. (4-64), we get

$$\Delta x \Delta p_x = 2p\lambda \qquad (4\text{-}66)$$

Substituting the de Broglie relationship $\lambda = h/p$, this becomes

$$\Delta x \Delta p_x = 2h \qquad (4\text{-}67)$$

Equation (4-67) tells us essentially the same as Eq. (4-63), i.e., the product of the uncertainties in the position and momentum of the electron is a finite quantity. The actual value of this quantity that defines the limit of our knowledge of the position and momentum is dependent on our definition of uncertainty. The important fact is that there is a limit and that its value is of the order of $h$.

We can now examine the problem of simultaneous measurement of position and momentum more closely. What happens if we try to locate the electron more precisely? We could do this by decreasing the width of the slit. But, according to Eq. (4-67), this would increase the uncertainty in our knowledge of the momentum. On the other hand, if we try to measure the momentum more precisely, we can do so only at the expense of the precision of measurement of the location.

The experiment we have just described leads us to suggest that the uncertainty principle is in accord with what we can experimentally observe. Of course, we could think of other experiments that might give us exact knowledge of the simultaneous position and momentum of the electron, and in fact, many ingenious proposals have been made. In all cases, it has been discovered that there is a finite value for the lower limit of the product of the uncertainties in the measurement of the position and momentum. This is a basic difference between classical and quantum mechanics. In classical mechanics it is assumed that with sufficiently accurate instruments it is possible to make precise measurements of position and momentum simultaneously. Quantum mechanics, on the other hand, insists that precise simultaneous determination of position and momentum is impossible. However fine our experiments, according to quantum mechanics, there is a limit to the accuracy of our simultaneous determination of these two quantities. The limitation is due to the dual nature that quantum mechanics accords to particles.

We have expressed the uncertainty principle in terms of position and momentum. There are, however, other pairs of variables that are subject to the uncertainty relationship. Energy and time constitute an important pair, and we can express the uncertainty principle in terms of these. By comparison with Eq. (4-63) we write

$$\Delta E \Delta t \geqslant \frac{\hbar}{2} \qquad (4\text{-}68)$$

Here the uncertainty in the time $\Delta t$ is the time associated with the measurement of the energy. For time-independent phenomena, $\Delta t \rightarrow \infty$, which means that $\Delta E \rightarrow 0$; i.e., the energy is precisely defined, but we have no knowledge of time. Equation (4-68) can be used to estimate the width of spectral lines (Exercise 4-4) since the life-times of excited states are sometimes very short (i.e., $\Delta t$ is small).

There is just one further point we should make: We would not expect to see the effects of the uncertainty principle in our macroscopic world. This is because the uncertainties are far too small to be detectable in our everyday dealings with macroscopic bodies.

## 4-13  THE HAMILTONIAN OPERATOR

In postulate 3 we introduced the *Hamiltonian operator*. What does this represent?

In classical mechanics the total energy of a system is given by *Hamilton's function H*:

$$H = T + U$$

$T$ is the total kinetic energy; $U$ is the total potential energy. We can write this in terms of the linear momentum and the coordinates of the system. If we do this for a single particle of mass $m$, $H$ becomes

$$H = \frac{1}{2m}(p_x^2 + p_y^2 + p_z^2) + U(x, y, z, t) \qquad (4\text{-}69)$$

$p_x, p_y$, and $p_z$ are the components of the momentum in the directions of the coordinate axes.

Now look back to postulate 2. This says that to construct the quantum mechanical operator from the classical expression we have to replace the momenta by, for example, $(\hbar/i)(\partial/\partial x)$ for $p_x$; we leave the coordinates as they are. Let us do this with the Hamiltonian function Eq. (4-69); we get

$$\hat{H} = \frac{1}{2m}\left[\left(\frac{\hbar}{i}\frac{\partial}{\partial x}\right)^2 + \left(\frac{\hbar}{i}\frac{\partial}{\partial y}\right)^2 + \left(\frac{\hbar}{i}\frac{\partial}{\partial z}\right)^2\right] + U(x, y, z, t)$$

i.e.,

$$\hat{H} = -\frac{\hbar^2}{2m}\left(\frac{\partial^2}{\partial x^2} + \frac{\partial^2}{\partial y^2} + \frac{\partial^2}{\partial z^2}\right) + U(x, y, z, t) \qquad (4\text{-}70)$$

Using the Laplacian operator we met before (cf. Section 4-2, postulate 2), we can write this as

$$\hat{H} = -\frac{\hbar^2}{2m}\nabla^2 + U \qquad (4\text{-}71)$$

We call the operator $\hat{H}$ the *Hamiltonian operator*. It is the operator associated with the energy of the system. Energy is the most important observable in chemistry, so the Hamiltonian operator is very important to us: we shall find it cropping up continually from now on.

## 4-14  SCHROEDINGER'S EQUATION FOR A STATIONARY STATE

We can now obtain from our postulates the most important equation we need in this book — Schroedinger's equation for a stationary state:

Look at postulate 4. It says that the value of an observable $a$ corresponding to an operator $\hat{A}$ must satisfy the eigenvalue equation

$$\hat{A}\Psi(q, t) = a\Psi(q, t)$$

$\Psi$ is the wavefunction for the system on which we made the measurement. If our observable is the total energy $E$, then the operator is the Hamiltonian operator $\hat{H}$. Therefore, we can write

$$\hat{H}\Psi(q, t) = E\Psi(q, t) \qquad (4\text{-}72)$$

where $E$ does not vary with time. We say that the system described by Eq. (4-72) is in a stationary state.

Now look at postulate 3. This says that the Hamiltonian operator satisfies the equation

$$\hat{H}\Psi(q, t) = -\frac{\hbar}{i}\frac{\partial}{\partial t}\Psi(q, t) \qquad (4\text{-}3)$$

Obviously, we must make Eqs. (4-72) and (4-3) consistent. We can do this by applying the method of separation of variables (cf. Section 3-2) to Eq. (4-3). We write

$$\Psi(q, t) = \Psi(q)\phi(t) \qquad (4\text{-}73)$$

Substitute this into Eq. (4-3); we get

$$\hat{H}\Psi(q, t) = -\frac{\hbar}{i}\Psi(q)\frac{d\phi(t)}{dt} \qquad (4\text{-}74)$$

Using Eq. (4-72), this becomes

$$E\Psi(q, t) = -\frac{\hbar}{i}\Psi(q)\frac{d\phi(t)}{dt} \qquad (4\text{-}75)$$

Divide by $\Psi(q, t)$:

$$E = -\frac{\hbar}{i}\frac{1}{\phi(t)}\frac{d\phi(t)}{dt} \qquad (4\text{-}76)$$

This is an ordinary differential equation. The solution is

$$\phi(t) = \exp\left(-\frac{iEt}{\hbar}\right) \qquad (4\text{-}77)$$

Substitute this into Eq. (4-73):

$$\Psi(q, t) = \Psi(q) \exp\left(-\frac{iEt}{\hbar}\right) \tag{4-78}$$

Now substitute Eq. (4-78) into Eq. (4-3); we get

$$\hat{H}\Psi(q) \exp\left(-\frac{iEt}{\hbar}\right) = -\frac{\hbar}{i}\frac{\partial}{\partial t}\left[\Psi(q) \exp\left(-\frac{iEt}{\hbar}\right)\right]$$

$$= E \exp\left(-\frac{iEt}{\hbar}\right)\Psi(q) \tag{4-79}$$

i.e.,

$$\hat{H}\Psi(q) = E\Psi(q) \tag{4-80}$$

This is the equation we set out to obtain: it is *Schroedinger's equation for a stationary state*. It is applicable when we are not dealing with time-dependent phenomena. Note that the wavefunction in this equation is now independent of time — it is a function of the spatial coordinates only.

If we consider a single particle of mass $m$, the Hamiltonian takes the form Eq. (4-71) with $U$ independent of time. Substitution of this into Eq. (4-80) gives

$$\left(-\frac{\hbar^2}{2m}\nabla^2 + U\right)\Psi(q) = E\Psi(q) \tag{4-81}$$

This equation should be familiar to us: it is the one we obtained in Section 2-3.

What have we done so far in our general approach to quantum mechanics? We have got back to where we were in Chapter 2, Eq. (2-14). There we derived the Schroedinger equation from a wave equation and by recognizing that matter shows wave-like properties; here we have derived the same equation from a very general set of postulates — postulates that will allow us to build up a quantum mechanical model of nature.

## SUMMARY

1. We put forward five postulates on which to base the quantum mechanical theory. We summarized these in Section 4-2, so we shall not do so again. Here we shall review the more important consequences of the postulates.

2. The normalization condition is

$$\int \Psi^*\Psi \, d\tau = 1 \tag{4-12}$$

3. The operators we use in quantum mechanics are *Hermitian*. Hermitian operators have real eigenvalues.

4. Different eigenfunctions $\Psi_1$ and $\Psi_2$ corresponding to different eigenvalues of a Hermitian operator are *orthogonal*. We write this condition

$$\int \Psi_1^* \Psi_2 d\tau = 0 \tag{4-19}$$

Orthogonality means that the functions are completely independent of each other.

5. The eigenfunctions of the operators we use in quantum mechanics form *complete orthonormal sets*. Any function with the same boundary conditions as a complete orthonormal set can be expanded as a linear combination of the members of the set.

6. The average value $\bar{a}$ that can be measured for an observable $a$ whose operator is $\hat{A}$ is given by

$$\bar{a} = \int \Psi^* \hat{A} \Psi d\tau \tag{4-41}$$

where $\Psi$ is the wavefunction for the system on which the measurement is made.

7. Simultaneous eigenstates of two operators $\hat{A}$ and $\hat{B}$ can exist only if the operators commute, i.e., $\hat{A}\hat{B} - \hat{B}\hat{A} = 0$. This means that we can measure two observables precisely and simultaneously only if their operators commute: this is in accord with the *uncertainty principle*.

8. The energy operator is the *Hamiltonian* $\hat{H}$. For a single particle of mass $m$ in a potential $U$, this takes the form

$$\hat{H} = -\frac{\hbar^2}{2m}\nabla^2 + U \tag{4-71}$$

From this we can obtain *Schroedinger's equation for a stationary state* of the particle:

$$\left(-\frac{\hbar^2}{2m}\nabla^2 + U\right)\Psi(q) = E\Psi(q) \tag{4-81}$$

## EXERCISES

**4-1**  Show that the operator associated with $p_x$ is Hermitian. (Hint: Use partial integration.)

**4-2**  Using Eq. (3-19), determine the average value for the momentum of a particle in a one-dimensional box of length $a$. Also, determine the average value for the square of the momentum.

**4-3**  Consider two eigenfunctions $\Psi_i$ and $\Psi_j$ of an operator $\hat{A}$. If the eigenvalues associated with these functions are different, show that the following integral is zero:

$$\int \Psi_i^* \hat{A} \Psi_j d\tau \tag{4-82}$$

**4-4**  The effect of the uncertainty principle is one reason for the broadening of spectral lines — the extent of this broadening can be illustrated by the following problem: Estimate the minimum width of a spectral line produced by a transition from the ground state of an atom (which in general has a long life-time) to an excited state with a life-time of $5 \times 10^{-9}$ sec.

**4-5**  Show that the operators for position of a particle, $\hat{x}\cdot$, $\hat{y}\cdot$, and $\hat{z}\cdot$, do not commute with the operator for its energy $\hat{H}$. What is the physical significance of this? i.e., what conclusion can we draw from this result regarding the measurement of the position and energy of the particle?

**4-6**  Show that the wavefunctions for a particle in a one-dimensional box (given by Eq. (3-23)) are orthonormal.

# 5

# The Treatment of Angular Momentum in Quantum Mechanics

## 5-1 INTRODUCTION

You should recall the concept of angular momentum from your elementary physics courses. Angular momentum, you will remember, is one of those very important properties that is conserved when a change takes place in an isolated system. Angular momentum is also important to us as chemists; it allows us to classify atomic states very conveniently; to a lesser extent it allows us to classify molecular states. Consequently, the very important field of spectroscopy makes considerable use of this concept. Apart from this, we cannot understand the detailed nature of matter — one of the most important problems in chemistry — without an understanding of angular momentum. It will be well worth our while, therefore, to find out how to deal with angular momentum in quantum theory.

## 5-2 OPERATORS FOR ANGULAR MOMENTUM

Angular momentum is an observable property of a system. Therefore, according to our postulates, it can be represented by an operator. How do we obtain this operator? We do so by simply applying the rules we previously laid down. First we find the classical expression for angular momentum in terms of the coordinates for the system and the linear momenta; then we convert this to a quantum mechanical operator by replacing each position coordinate $q_k$ by $q_k \cdot$, and each linear momentum $p_k$ by $(\hbar/i)\,(\partial/\partial q_k)$. Let us do this.

In classical mechanics angular momentum is a vector quantity, i.e., it has direction in space as well as magnitude. If a mass $m$ moves about a fixed point $P$ the angular momentum $\mathbf{M}$ is defined as

$$\mathbf{M} = \mathbf{r} \times m\mathbf{v}$$

Here $\mathbf{r}$ is a vector whose magnitude is the distance of the mass from $P$; $\mathbf{v}$ is another vector—the linear velocity of the mass. Since $m\mathbf{v}$ is the linear momentum $\mathbf{p}$, we can write

$$\mathbf{M} = \mathbf{r} \times \mathbf{p}$$

In mathematics we say that this type of product of two vectors is a *vector product* (sometimes it is called a *cross product*). Before we continue it will be instructive to examine the nature of the vector product.

## Vector Product in Vector Analysis

For two vectors $\mathbf{A}$ and $\mathbf{B}$ we write the vector product $\mathbf{C}$ as

$$\mathbf{C} = \mathbf{A} \times \mathbf{B}$$

By definition this is a vector with magnitude given by

$$|\mathbf{C}| = |\mathbf{A} \times \mathbf{B}| = |\mathbf{A}||\mathbf{B}| \sin \theta \tag{5-1}$$

$\theta$ is the angle between $\mathbf{A}$ and $\mathbf{B}$. The direction of $\mathbf{C}$ is perpendicular to the plane containing $\mathbf{A}$ and $\mathbf{B}$; it points in the direction of advance of a right-hand screw rotated from $\mathbf{A}$ to $\mathbf{B}$. Figure 5-1 should clarify this definition.

Since the vector product is itself a vector, it must have magnitude and direction; consequently, we can express it in terms of components along the coordinate axes. We shall now find general expressions for these components.

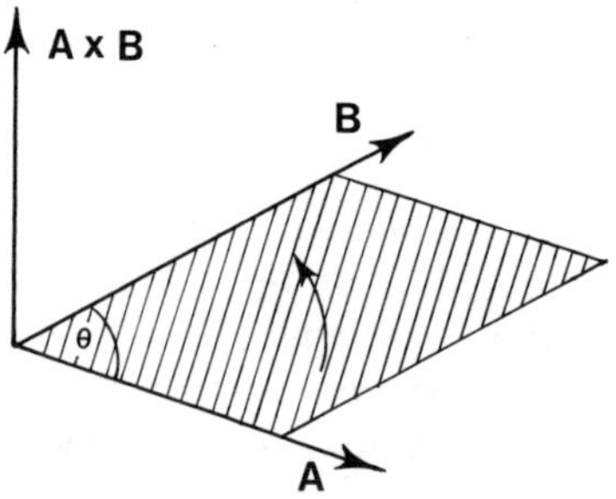

Our vectors $\mathbf{A}$ and $\mathbf{B}$ can be written

$$\mathbf{A} = A_x\mathbf{i} + A_y\mathbf{j} + A_z\mathbf{k}$$

$$\mathbf{B} = B_x\mathbf{i} + B_y\mathbf{j} + B_z\mathbf{k}$$

$\mathbf{i}$, $\mathbf{j}$, and $\mathbf{k}$ are mutually perpendicular, unit vectors parallel to the coordinate axes; $A_x$, $A_y$, $A_z$, $B_x$, $B_y$, and $B_z$ are the components of $\mathbf{A}$ and $\mathbf{B}$ parallel to the axes. Our vector is therefore

$$\begin{aligned}
\mathbf{A} \times \mathbf{B} &= (A_x\mathbf{i} + A_y\mathbf{j} + A_z\mathbf{k}) \times (B_x\mathbf{i} + B_y\mathbf{j} + B_z\mathbf{k}) \\
&= A_xB_x\mathbf{i} \times \mathbf{i} + A_xB_y\mathbf{i} \times \mathbf{j} + A_xB_z\mathbf{i} \times \mathbf{k} \\
&\quad + A_yB_x\mathbf{j} \times \mathbf{i} + A_yB_y\mathbf{j} \times \mathbf{j} + A_yB_z\mathbf{j} \times \mathbf{k} \\
&\quad + A_zB_x\mathbf{k} \times \mathbf{i} + A_zB_y\mathbf{k} \times \mathbf{j} + A_zB_z\mathbf{k} \times \mathbf{k} \\
&= (A_yB_z - A_zB_y)\mathbf{i} + (A_zB_x - A_xB_z)\mathbf{j} \\
&\quad + (A_xB_y - A_yB_x)\mathbf{k}
\end{aligned}$$

**Fig. 5-1** Diagrammatic representation of the vector product. $\mathbf{A} \times \mathbf{B}$ is perpendicular to the plane containing $\mathbf{A}$ and $\mathbf{B}$.

We can write the last step because the vector products of the unit vectors are zero, i.e.,

$$\mathbf{i}\times\mathbf{i}=\mathbf{j}\times\mathbf{j}=\mathbf{k}\times\mathbf{k}=0 \quad (\text{because } \sin\theta \text{ in Eq. (5-1)}=0)$$

Also, since the unit vectors are mutually perpendicular, Eq. (5-1) gives

$$\mathbf{i}\times\mathbf{j}=\mathbf{k}; \quad \mathbf{j}\times\mathbf{i}=-\mathbf{k}; \quad \mathbf{j}\times\mathbf{k}=\mathbf{i}$$
$$\mathbf{k}\times\mathbf{j}=-\mathbf{i}; \quad \mathbf{k}\times\mathbf{i}=\mathbf{j}; \quad \mathbf{i}\times\mathbf{k}=-\mathbf{j}$$

We have called the new vector resulting from the vector product $\mathbf{C}$. The components of $\mathbf{C}$ will therefore be

$$C_x = A_y B_z - A_z B_y$$
$$C_y = A_z B_x - A_x B_z$$
$$C_z = A_x B_y - A_y B_x$$

Now let us apply what we have learned here to our angular momentum vector. Let the components of $\mathbf{r}$ be $x$, $y$, and $z$. We can write for the components of $\mathbf{M}$

$$M_x = y p_z - z p_y \tag{5-2a}$$
$$M_y = z p_x - x p_z \tag{5-2b}$$
$$M_z = x p_y - y p_x \tag{5-2c}$$

We now have the angular momentum expressed classically in the form we want it. Applying our rules for conversion to quantum mechanical operators (replace $p_x$ by $(\hbar/i)(\partial/\partial x)$, etc.; leave the coordinates alone), we immediately get

$$\hat{M}_x = \frac{\hbar}{i}\left(y\frac{\partial}{\partial z} - z\frac{\partial}{\partial y}\right) \tag{5-3a}$$
$$\hat{M}_y = \frac{\hbar}{i}\left(z\frac{\partial}{\partial x} - x\frac{\partial}{\partial z}\right) \tag{5-3b}$$
$$\hat{M}_z = \frac{\hbar}{i}\left(x\frac{\partial}{\partial y} - y\frac{\partial}{\partial x}\right) \tag{5-3c}$$

Here we can obtain an interesting and very important result. By Section 4-12, we can know two observables in a system precisely only if their operators commute. Let us see whether our operators for the angular momentum components commute:

$$\hat{M}_x\hat{M}_y - \hat{M}_y\hat{M}_x = -\hbar^2\left(y\frac{\partial}{\partial z} - z\frac{\partial}{\partial y}\right)\left(z\frac{\partial}{\partial x} - x\frac{\partial}{\partial z}\right)$$
$$+ \hbar^2\left(z\frac{\partial}{\partial x} - x\frac{\partial}{\partial z}\right)\left(y\frac{\partial}{\partial z} - z\frac{\partial}{\partial y}\right)$$

$$= \hbar^2 \left( -yz\frac{\partial^2}{\partial z \partial x} - y\frac{\partial}{\partial x} + yx\frac{\partial^2}{\partial z^2} + z^2\frac{\partial^2}{\partial y \partial x} - zx\frac{\partial^2}{\partial y \partial z} \right.$$

$$\left. + zy\frac{\partial^2}{\partial x \partial z} - z^2\frac{\partial^2}{\partial x \partial y} - xy\frac{\partial^2}{\partial z^2} + xz\frac{\partial^2}{\partial z \partial y} + x\frac{\partial}{\partial y} \right)$$

$$= \hbar^2 \left( x\frac{\partial}{\partial y} - y\frac{\partial}{\partial x} \right)$$

$$= i\hbar \hat{M}_z \qquad \text{(by Eq. (5-3c))} \tag{5-4}$$

By noting the symmetry of the problem we can write down the other commutation relations. The complete set is

$$\hat{M}_x\hat{M}_y - \hat{M}_y\hat{M}_x = i\hbar\hat{M}_z \tag{5-5a}$$

$$\hat{M}_y\hat{M}_z - \hat{M}_z\hat{M}_y = i\hbar\hat{M}_x \tag{5-5b}$$

$$\hat{M}_z\hat{M}_x - \hat{M}_x\hat{M}_z = i\hbar\hat{M}_y \tag{5-5c}$$

The important point for us to note here is that these operators do not commute. What does this mean? It means that we cannot know precisely more than one of the components of angular momentum for a system. We can know one component and the average values of the others, but this is the best we can do.

Though the operators for the components of angular momentum do not commute, there is an operator that does commute with the components: this is the operator for the square of the angular momentum, $\hat{M}^2$. Let us prove this by applying our rules for finding operators, then doing the multiplications. The square of the angular momentum is a scalar; it is given by the scalar product of **M** with itself:

$$\mathbf{M} \cdot \mathbf{M} = M^2 = M_x^2 + M_y^2 + M_z^2 \tag{5-6}$$

Using this last equation, we write

$$\hat{M}^2\hat{M}_z - \hat{M}_z\hat{M}^2 = (\hat{M}_x^2 + \hat{M}_y^2 + \hat{M}_z^2)\hat{M}_z - \hat{M}_z(\hat{M}_x^2 + \hat{M}_y^2 + \hat{M}_z^2)$$

$$= \hat{M}_x^2\hat{M}_z - \hat{M}_z\hat{M}_x^2 + \hat{M}_y^2\hat{M}_z - \hat{M}_z\hat{M}_y^2 \tag{5-7}$$

$$\text{(because } \hat{M}_z^2\hat{M}_z - \hat{M}_z\hat{M}_z^2 = \hat{M}_z^3 - \hat{M}_z^3 = 0)$$

(Note that we must not change the order of multiplication of the operators in Eq. (5-7): we could do so only if they commuted — we do not know whether they do or not.) Now we write for the first two terms on the right side of Eq. (5-7)

$$\hat{M}_x{}^2\hat{M}_z - \hat{M}_z\hat{M}_x{}^2 = \hat{M}_x\hat{M}_x\hat{M}_z - \hat{M}_z\hat{M}_x\hat{M}_x$$
$$= \hat{M}_x(\hat{M}_x\hat{M}_z - \hat{M}_z\hat{M}_x) - (\hat{M}_z\hat{M}_x - \hat{M}_x\hat{M}_z)\hat{M}_x$$
$$= \hat{M}_x \cdot -i\hbar\hat{M}_y - i\hbar\hat{M}_y\hat{M}_x \qquad \text{(by Eq. (5-5c))}$$
$$= -i\hbar(\hat{M}_x\hat{M}_y + \hat{M}_y\hat{M}_x) \qquad\qquad (5\text{-}8)$$

Similarly, we can write for the third and fourth terms on the right side of Eq. (5-7)

$$\hat{M}_y{}^2\hat{M}_z - \hat{M}_z\hat{M}_y{}^2 = i\hbar(\hat{M}_y\hat{M}_x + \hat{M}_x\hat{M}_y) \qquad (5\text{-}9)$$

Add Eqs. (5-8) and (5-9); we get

$$\hat{M}^2\hat{M}_z - \hat{M}_z\hat{M}^2 = 0 \qquad\qquad (5\text{-}10\text{a})$$

By symmetry we can write

$$\hat{M}^2\hat{M}_y - \hat{M}_y\hat{M}^2 = 0 \qquad\qquad (5\text{-}10\text{b})$$
$$\hat{M}^2\hat{M}_x - \hat{M}_x\hat{M}^2 = 0 \qquad\qquad (5\text{-}10\text{c})$$

i.e., $\hat{M}^2$ and the operators for the components of the angular momentum *do* commute. This means that we can measure $M^2$ precisely (and, therefore, the magnitude of the total angular momentum, which is equal to $|\mathbf{M}| = \sqrt{M^2} = \sqrt{M_x{}^2 + M_y{}^2 + M_z{}^2}$) and a component in one direction; we cannot also measure a second component, since this would fix the value of the third component (because we know the value of the total angular momentum). Our commutation rules (and the uncertainty principle) tell us that we cannot know $M_x$, $M_y$, and $M_z$ simultaneously.

## 5-3  EIGENVALUES OF THE ANGULAR MOMENTUM OPERATORS

Now that we have examined the angular momentum operators in some detail, the obvious question to ask is: What are the eigenstates of these operators? i.e., what eigenfunctions and eigenvalues do they have?

We shall not examine the form of the eigenfunctions here because we do this in the next chapter when we consider the quantum mechanical treatment of the rigid rotator. Here we shall just consider the eigenvalues. We can obtain these in two ways: 1. We can consider the forms of the operators and solve the eigenvalue equations. 2. We can define the operators according to the commutation relations Eqs. (5-5) and (5-10) and show that the eigenvalues are a logical consequence of these relations. We shall choose the second method, though we see that the commutation relations result from the forms of the operators; consequently, both methods are entirely equivalent. The mathematics is, perhaps, tedious. However, it is straightforward enough if we understand the properties of operators. It is well worth our while working through it because the results are so fundamental to quantum chemistry — the mathematics shows how two of the basic quantum numbers evolve from the logical structure of quantum theory.

First, recall postulate 4. This says that any operator in quantum mechanics that represents an observable must satisfy an eigenvalue equation of the form

$$\hat{A}\Psi_\lambda = a_\lambda \Psi_\lambda$$

Next, recall that if operators commute, they must possess simultaneous eigenfunctions; in other words, their eigenfunctions are the same. Therefore, we can write

$$\hat{M}^2 Y_{l,m} = k_l Y_{l,m} \tag{5-11}$$

$$\hat{M}_z Y_{l,m} = k_m Y_{l,m} \tag{5-12}$$

$k_l$ and $k_m$ are constants — they are the eigenvalues of the operators $\hat{M}^2$ and $\hat{M}_z$. (We could have chosen $\hat{M}_x$ or $\hat{M}_y$ instead of $\hat{M}_z$ since there is no preferred direction in space in the absence of external fields. However, it is conventional to use $\hat{M}_z$ in this discussion.)

Probably, you have wondered why we choose to call the eigenfunctions $Y_{l,m}$ instead of something simpler. The reason is that these functions turn out to be functions that are well-known to mathematicians — in fact, they were discovered long before quantum theory was formulated. We call them *spherical harmonics* and use the mathematically conventional symbol $Y_{l,m}$. When we consider the quantum mechanical treatment of the rigid rotator we shall examine them in more detail. In the meantime we shall not bother ourselves any further with their form — we shall just consider the eigenvalues associated with them.

To return to the problem: Operate on both sides of Eq. (5-12) with $\hat{M}_z$; we get

$$\hat{M}_z^2 Y_{l,m} = k_m \hat{M}_z Y_{l,m} = k_m^2 Y_{l,m} \tag{5-13}$$

Now

$$\hat{M}_z^2 = \hat{M}^2 - (\hat{M}_x^2 + \hat{M}_y^2) \qquad \text{(by Eq. (5-6))}$$

Therefore,

$$\hat{M}_z^2 Y_{l,m} = [\hat{M}^2 - (\hat{M}_x^2 + \hat{M}_y^2)] Y_{l,m}$$

$$= k_l Y_{l,m} - (\hat{M}_x^2 + \hat{M}_y^2) Y_{l,m} \tag{5-14}$$

Substitute this into Eq. (5-13):

$$k_l Y_{l,m} - (\hat{M}_x^2 + \hat{M}_y^2) Y_{1,m} = k_m^2 Y_{l,m} \qquad (5\text{-}15)$$

Therefore,

$$(\hat{M}_x^2 + \hat{M}_y^2) Y_{l,m} = (k_l - k_m^2) Y_{l,m} \qquad (5\text{-}16)$$

$(\hat{M}_x^2 + \hat{M}_y^2)$ is an operator for an observable $(M_x^2 + M_y^2)$. Now this observable cannot be negative; i.e., the eigenvalue of $(\hat{M}_x^2 + \hat{M}_y^2)$ cannot be negative. According to Eq. (5-16), the eigenvalue is $(k_l - k_m^2)$, so

$$k_l \geqslant k_m^2 \qquad (5\text{-}17)$$

This is an important result, which we shall use later in this discussion. In the meantime, consider the operator expression

$$\hat{M}_z(\hat{M}_x + i\hat{M}_y) = \hat{M}_z\hat{M}_x + i\hat{M}_z\hat{M}_y \qquad (5\text{-}18)$$

The reason for considering this will become clear soon. We can rearrange the right side of Eq. (5-18) with the help of the commutation relations between the operators for the components of angular momentum, Eq. (5-5). The expression becomes

$$i\hbar\hat{M}_y + \hat{M}_x\hat{M}_z + i\hat{M}_y\hat{M}_z + \hbar\hat{M}_x = (\hat{M}_x + i\hat{M}_y)(\hat{M}_z + \hbar) \qquad (5\text{-}19)$$

Similarly, we can show that

$$\hat{M}_z(\hat{M}_x - i\hat{M}_y) = (\hat{M}_x - i\hat{M}_y)(\hat{M}_z - \hbar) \qquad (5\text{-}20)$$

Now let both sides of Eqs. (5-19) and (5-20) operate on $Y_{l,m}$, the eigenfunctions of the angular momentum operators; we get

$$\hat{M}_z(\hat{M}_x + i\hat{M}_y) Y_{l,m} = (\hat{M}_x + i\hat{M}_y)(\hat{M}_z + \hbar) Y_{l,m} \qquad (5\text{-}21)$$

$$\hat{M}_z(\hat{M}_x - i\hat{M}_y) Y_{l,m} = (\hat{M}_x - i\hat{M}_y)(\hat{M}_z - \hbar) Y_{l,m} \qquad (5\text{-}22)$$

Using Eq. (5-12), we can write Eqs. (5-21) and (5-22)

$$\hat{M}_z(\hat{M}_x + i\hat{M}_y) Y_{l,m} = (k_m + \hbar)(\hat{M}_x + i\hat{M}_y) Y_{l,m} \qquad (5\text{-}23)$$

$$\hat{M}_z(\hat{M}_x - i\hat{M}_y) Y_{l,m} = (k_m - \hbar)(\hat{M}_x - i\hat{M}_y) Y_{l,m} \qquad (5\text{-}24)$$

These equations define eigenfunctions of $\hat{M}_z$. The functions are $(\hat{M}_x + i\hat{M}_y) Y_{l,m}$ and $(\hat{M}_x - i\hat{M}_y) Y_{l,m}$; the corresponding eigenvalues are $(k_m + \hbar)$ and $(k_m - \hbar)$. Since $\hat{M}^2$ commutes with $\hat{M}_x$ and $\hat{M}_y$, these eigenfunctions must correspond to the eigenvalue $k_l$ of $\hat{M}^2$. Therefore, we can say that for the eigenvalue $k_l$ of $\hat{M}^2$ there are two eigenvalues of $\hat{M}_z$; these are $(k_m + \hbar)$ and $(k_m - \hbar)$. But these are not all the eigenvalues of $\hat{M}_z$ that correspond to the single eigenvalue $k_l$ of $\hat{M}^2$; for if we consider the operators $(\hat{M}_x + i\hat{M}_y)^2$ and $(\hat{M}_x - i\hat{M}_y)^2$ we can show that

$$\hat{M}_z(\hat{M}_x + i\hat{M}_y)^2 Y_{l,m} = (k_m + 2\hbar)(\hat{M}_x + i\hat{M}_y)^2 Y_{l,m} \qquad (5\text{-}25)$$

$$\hat{M}_z(\hat{M}_x - i\hat{M}_y)^2 Y_{l,m} = (k_m - 2\hbar)(\hat{M}_x - i\hat{M}_y)^2 Y_{l,m} \qquad (5\text{-}26)$$

Also,

$$\hat{M}_z(\hat{M}_x + i\hat{M}_y)^3 Y_{l,m} = (k_m + 3\hbar)(\hat{M}_x + i\hat{M}_y)^3 Y_{l,m} \qquad (5\text{-}27)$$

$$\hat{M}_z(\hat{M}_x - i\hat{M}_y)^3 Y_{l,m} = (k_m - 3\hbar)(\hat{M}_x - i\hat{M}_y)^3 Y_{l,m} \quad \text{etc.} \qquad (5\text{-}28)$$

We shall not work through the details of obtaining these last equations. The mathematics uses the commutation relations for the angular momentum operators and is similar to that used to obtain Eqs. (5-23) and (5-24).

We can conclude from this discussion that for each eigenvalue $k_l$ of $\hat{M}^2$ there is a series of eigenvalues of $\hat{M}_z$; this series is

$$\ldots, k_m - 2\hbar, k_m - \hbar, k_m, k_m + \hbar, k_m + 2\hbar, \ldots \qquad (5\text{-}29)$$

According to Eq. (5-17), $k_1 \geqslant k_m{}^2$. Therefore, the series must terminate in both directions.

We call the operators $(\hat{M}_x + i\hat{M}_y)$ and $(\hat{M}_x - i\hat{M}_y)$ that generated the eigenfunctions corresponding to these eigenvalues of $\hat{M}_z$ *shift operators*. The first is called the *raising operator*; the second is called the *lowering operator*. Each time we operate on the function $Y_{l,m}$ with $(\hat{M}_x + i\hat{M}_y)$ we raise the eigenvalue by $\hbar$; each time we operate with $(\hat{M}_x - i\hat{M}_y)$ we lower the eigenvalue by $\hbar$. We can now use these operators to determine the relationship between $k_1$ and $k_m$.

Let the lowest eigenvalue of the series Eq. (5-29) be $k'_m$; let the highest be $k''_m$. Let the corresponding eigenfunctions be $Y'_{l,m}$ and $Y''_{l,m}$. We can write

$$\hat{M}_z Y'_{l,m} = k'_m Y'_{l,m} \qquad (5\text{-}30)$$

$$\hat{M}_z Y''_{l,m} = k''_m Y''_{l,m} \qquad (5\text{-}31)$$

Now $(\hat{M}_x + i\hat{M}_y)$ should generate a function with a higher eigenvalue than $Y''_{l,m}$. But this is not possible because, by definition, $k''_m$ is the highest eigenvalue. Therefore, we must have

$$(\hat{M}_x + i\hat{M}_y)Y''_{l,m} = 0 \qquad (5\text{-}32)$$

Also $(\hat{M}_x - i\hat{M}_y)$ should generate a function with a lower eigenvalue than $k'_m$. Therefore,

$$(\hat{M}_x - i\hat{M}_y)Y'_{l,m} = 0 \qquad (5\text{-}33)$$

Operate on Eq. (5-32) with $(\hat{M}_x - i\hat{M}_y)$; we get

$$(\hat{M}_x - i\hat{M}_y)(\hat{M}_x + i\hat{M}_y)Y''_{l,m} = 0 \qquad (5\text{-}34)$$

$$(\hat{M}_x{}^2 + \hat{M}_y{}^2 + i\hat{M}_x\hat{M}_y - i\hat{M}_y\hat{M}_x)Y''_{l,m} = 0 \qquad (5\text{-}35)$$

Using Eqs. (5-6) and (5-5a), this becomes

$$(\hat{M}^2 - \hat{M}_z{}^2 - \hbar\hat{M}_z)Y''_{l,m} = 0 \qquad (5\text{-}36)$$

Remembering that $\hat{M}^2$ and $\hat{M}_z$ commute, and using Eqs. (5-11) and (5-31), we get

$$(k_l - k_m''^2 - \hbar k_m'')Y''_{l,m} = 0 \qquad (5\text{-}37)$$

Since $Y''_{l,m}$ is not equal to zero

$$k_l = k_m''^2 + \hbar k_m \qquad (5\text{-}38)$$

Now go back to Eq. (5-33) and operate on it with $(\hat{M}_x + i\hat{M}_y)$; we get in a similar way

$$k_l = k_m'^2 - \hbar k_m' \qquad (5\text{-}39)$$

Remembering that $k_m'' > k_m'$, we see that Eqs. (5-38) and (5-39) can be consistent only if $k_m'' = -k_m'$. Also, the series Eq. (5-29) shows us that $k_m''$ must be greater than $k_m'$ by an integral multiple of $\hbar$; therefore,

$$k_m'' = k_m' + n\hbar \qquad (5\text{-}40)$$

Consequently, we can write

$$-k_m' = k_m' + n\hbar$$

$$-k_m' = \frac{n}{2}\hbar \qquad (5\text{-}41)$$

Now the value of $k_m'$ depends on the value of $l$ only. We have not so far placed any particular significance on $l$; there is no reason, therefore, why we should not put $l = n/2$ and write Eq. (5-41) as

$$-k_m' = l\hbar \qquad (5\text{-}42)$$

$l$ is integral or half integral. Substitute Eq. (5-42) into Eq. (5-39); we get

$$k_l = l(l+1)\hbar^2 \qquad (5\text{-}43)$$

$k_m$ can take a series of values, which are only dependent on $k_l$. We remember that $k_m''$, the highest member of the series, is equal to the negative of $k_m'$, the lowest member; also, we refer to Eq. (5-42) and write this series

$$l\hbar, (l-1)\hbar, (l-2)\hbar, \ldots, -(l-2)\hbar, -(l-1)\hbar, -l\hbar \qquad (5\text{-}44)$$

In general we can write for $k_m$

$$k_m = m\hbar \tag{5-45}$$

where $m$ can take the $(2l+1)$ values with $-l \leqslant m \leqslant +l$. $m$ is an integer when $l$ is an integer; when $l$ is half integral, $m$ is also half integral.

We can now go back to Eqs. (5-11) and (5-12), the equations that define the eigenfunctions and eigenvalues of the angular momentum operators, and substitute for $k_l$ and $k_m$ from Eqs. (5-43) and (5-45); we get

$$\hat{M}^2 Y_{l,m} = l(l+1)\hbar^2 Y_{l,m} \tag{5-46}$$

$$\hat{M}_z Y_{l,m} = m\hbar Y_{l,m} \tag{5-47}$$

We said a few sentences ago that $l$ can be integral or half integral. For orbital angular momentum it turns out that $l$ is integral, i.e., $l = 0, 1, 2, 3,$ .... We could prove this by examining the form of the eigenfunctions of the angular momentum operators. But we have done enough mathematics for the moment: we shall leave the consideration of the eigenfunctions for the next chapter. In the meantime, let us briefly consider the results of this section; we shall also consider some of the immediate implications.

First we see that if $M^2$ takes the value $\hbar^2[l(l+1)]$ (cf. Eq. (5-46)), then the magnitude of the orbital angular momentum is

$$|\mathbf{M}| = \sqrt{M^2} = \hbar[l(l+1)]^{1/2} \tag{5-48}$$

However, the angular momentum is a vector. Our commutation rules (and the uncertainty principle) tell us that we cannot know the direction of this vector in space. We can know the magnitude and the value of one component, for example $M_z$, since $\hat{M}^2$ and $\hat{M}_z$ commute. But note that the only possible observable values of a component of angular momentum about a given axis are $m\hbar$ (cf. Eq. (5-45)). The maximum value of $m$ is $l$. Therefore, the maximum value of the component in any given direction is $l\hbar$—this is less than the total angular momentum given by Eq. (5-48). This, of course, is as it must be: if a component of angular momentum could be equal to the total angular momentum, then the direction would be completely specified since the other two components would have to be zero.

To conclude this section we see that consideration of orbital angular momentum has given us two quantum numbers, $l$ and $m$. We shall call them *orbital angular momentum quantum numbers*. These are the ones you have met before in your introductory physical chemistry courses; then they might have been called azimuthal and magnetic quantum numbers, these latter names being a throw-back to the old Bohr theory. The quantum numbers $l$ and $m$ therefore evolve naturally from the consideration of orbital angular momentum. We shall see later that the principle quantum number $n$ is associated with the energy operator. The fourth quantum number comes into the theory when we consider spin — this we shall now do.

## 5-4  TREATMENT OF SPIN

Some elementary particles have intrinsic angular momentum. For example, an electron in an atom, as well as having orbital angular momentum has this intrinsic angular momentum. We recognize the existence of this by saying that the electron spins; this, however, is merely a convenient way for us to picture the possession of intrinsic angular momentum.

If we wish to deal with spin in a quantum mechanical way, we naturally look for an operator to represent it. According to our rules for finding operators, we should look for a classical expression in terms of space coordinates and linear momenta, then apply the rules for converting this expression to a quantum mechanical operator. Unfortunately, we cannot do this for spin because we cannot find a classical expression to start from: spin is essentially a quantum mechanical phenomenon; it has no classical analog. This forces us to look for a different method of dealing with it. There are several methods available; we shall use a method that was developed by Pauli[4].

We assume that the angular momentum described by spin can be dealt with, at least initially, in the same way as orbital angular momentum. This, admittedly, is a rather arbitrary assumption; and later we shall have to make further assumptions. However, these assumptions are very well justified by the success they give when we try to explain certain experimental observations. Examples of these observations are the Stern–Gerlach experiment and the splitting of the spectral lines for the alkali metals into doublets. These experimental results cannot be explained with classical theory. The assumptions we are about to make do allow them to be explained with quantum theory.

For a start we define operators that are analogous to $\hat{M}^2$, $\hat{M}_x$, $\hat{M}_y$, and $\hat{M}_z$. We give them the symbol $\hat{S}$, so the corresponding spin operators will be $\hat{S}^2$, $\hat{S}_x$, $\hat{S}_y$, and $\hat{S}_z$. Now we assume that these operators obey the same commutation rules as those for orbital angular momentum. These may therefore be written (cf. Eqs. (5-5) and (5-10))

$$\hat{S}_x\hat{S}_y - \hat{S}_y\hat{S}_x = i\hbar\hat{S}_z \qquad (5\text{-}49\text{a})$$

$$\hat{S}_y\hat{S}_z - \hat{S}_z\hat{S}_y = i\hbar\hat{S}_x \qquad (5\text{-}49\text{b})$$

$$\hat{S}_z\hat{S}_x = \hat{S}_x\hat{S}_z = i\hbar\hat{S}_y \tag{5-49c}$$

i.e., the operators for the components do not commute.

$$\hat{S}_x\hat{S}^2 - \hat{S}^2\hat{S}_x = 0 \tag{5-50a}$$

$$\hat{S}_y\hat{S}^2 - \hat{S}^2\hat{S}_y = 0 \tag{5-50b}$$

$$\hat{S}_z\hat{S}^2 - \hat{S}^2\hat{S}_z = 0 \tag{5-50c}$$

i.e., $\hat{S}^2$ does commute with the operators for the components.

Now we must introduce a difference between orbital angular momentum operators and spin operators. The operators for orbital angular momentum have many possible eigenstates associated with them; from experimental evidence, such as the Stern–Gerlach experiment, it appears that the spin operator for a single electron can have only two eigenstates associated with it. Therefore, we assume that for a single electron there are only two eigenfunctions of $\hat{S}^2$ and $\hat{S}_z$: it is conventional to call these $\alpha$ and $\beta$. (We use $\hat{S}_z$ here though, of course, we could just as well have used $\hat{S}_x$ or $\hat{S}_y$.)

At this point we recall our discussion of the eigenvalues of the operators associated with orbital angular momentum (Section 5-3). There we decided that $l$ could be integral or half integral; also, if $l$ is integral, then $m$ is integral, and if $l$ is half integral, then $m$ is half integral. These results were perfectly general for operators obeying the same commutation relations as the orbital angular momentum operators. The spin operators obey these rules; therefore, we can say something about their eigenvalues. But, so that our results will agree with experimental evidence—as they must— we let the quantum number corresponding to $l$ in the orbital angular momentum theory be $\frac{1}{2}$. If we then call the quantum number corresponding to $m$ in the orbital angular momentum theory $m_s$, $m_s$ takes the values $+\frac{1}{2}$ and $-\frac{1}{2}$. This gives us two spin eigenstates for a single electron.

Referring back to our analogy with orbital angular momentum, we can write (cf. Eqs. (5-46) and (5-47))

$$\hat{S}^2\alpha = \frac{1}{2}(\frac{1}{2}+1)\hbar^2\alpha = \frac{3}{4}\hbar^2\alpha \tag{5-51}$$

$$\hat{S}^2\beta = \frac{1}{2}(\frac{1}{2}+1)\hbar^2\beta = \frac{3}{4}\hbar^2\beta \tag{5-52}$$

$$\hat{S}_z\alpha = \frac{1}{2}\hbar\alpha \tag{5-53}$$

$$\hat{S}_z\beta = -\frac{1}{2}\hbar\beta \tag{5-54}$$

## SUMMARY

1. We found the operators for the components of angular momentum ($\hat{M}_x$, $\hat{M}_y$, and $\hat{M}_z$) by applying our rules for determining quantum mech-

anical operators. These operators are

$$\hat{M}_x = \frac{\hbar}{i}\left(y\frac{\partial}{\partial z} - z\frac{\partial}{\partial y}\right)$$    (5-3a)

$$\hat{M}_y = \frac{\hbar}{i}\left(z\frac{\partial}{\partial x} - x\frac{\partial}{\partial z}\right)$$    (5-3b)

$$\hat{M}_z = \frac{\hbar}{i}\left(x\frac{\partial}{\partial y} - y\frac{\partial}{\partial x}\right)$$    (5-3c)

2. We showed that these operators do not commute with each other; but they do commute individually with $\hat{M}^2$, the operator for the square of the total angular momentum.

3. We examined the eigenvalues of the angular momentum operators. Two quantum numbers, $l$ and $m$, resulted from this. $l$ is associated with $\hat{M}^2$; $m$ is associated with $\hat{M}_z$. $l$ can take the values $0, 1, 2, 3, \ldots$ ; $m$ then takes the integral values $-l \leqslant m \leqslant +l$.

4. Spin is a convenient way of picturing the intrinsic angular momentum of the electron. Formally, we deal with this as we dealt with orbital angular momentum. The difference is that there are only two eigenstates for the spin operator associated with a single electron. These eigenstates are labeled with the spin quantum numbers $m_s = \frac{1}{2}$, or $m_s = -\frac{1}{2}$.

## EXERCISES

**5-1**  Show that the operators for the components of angular momentum, as given in Eq. (5-3), are Hermitian.

**5-2**  It is often convenient to express the operators for angular momentum in polar coordinate form. By using the relationship between cartesian and polar coordinates, as given in Eq. (6-1), show that $\hat{M}_x, \hat{M}_y$, and $\hat{M}_z$ are given by

$$\hat{M}_x = \frac{\hbar}{i}\left(-\sin\phi\,\frac{\partial}{\partial\theta} - \cot\theta\cos\phi\,\frac{\partial}{\partial\phi}\right)$$

$$\hat{M}_y = \frac{\hbar}{i}\left(\cos\phi\,\frac{\partial}{\partial\theta} - \cot\theta\sin\phi\,\frac{\partial}{\partial\phi}\right)$$

$$\hat{M}_z = \frac{\hbar}{i}\frac{\partial}{\partial\phi}$$

Further, show that $\hat{M}^2$ is given by

$$\hat{M}^2 = -\hbar^2\left[\frac{1}{\sin\theta}\frac{\partial}{\partial\theta}\left(\sin\theta\,\frac{\partial}{\partial\theta}\right) + \frac{1}{\sin^2\theta}\frac{\partial^2}{\partial\phi^2}\right]$$

**5-3**  If we denote the raising operator by $\hat{M}_+$ and the lowering operator by $\hat{M}_-$, show that

$$\hat{M}^2 = \hat{M}_+\hat{M}_- + \hat{M}_z^2 - \hbar\hat{M}_z$$
$$= \hat{M}_-\hat{M}_+ + \hat{M}_z^2 + \hbar\hat{M}_z$$

**5-4**  Verify Eqs. (5-25) and (5-26).

**5-5**  By definition, the spin functions $\alpha$ and $\beta$ are orthogonal and normalized. Express these facts mathematically.

# 6

# Examples of Solutions of Schroedinger Equations

## 6-1 INTRODUCTION

In Chapter 3 we solved the Schroedinger equation for a very simple system—the particle-in-a-box. We are now in a position to solve some rather less simple problems. We choose three problems of interest to chemists—the rigid rotator, the harmonic oscillator and the hydrogen atom. These, of course, are still very simple systems by chemical standards. However, they give us some idea of what is involved in trying to solve Schroedinger equations for actual systems: they indicate the complexity of problems of interest to chemists.

## 6-2 THE RIGID ROTATOR

Our interest in the rigid rotator is that it approximates a diatomic molecule. Also, the mathematics involved in solving this problem is of considerable interest in other quantum mechanical problems. What we are going to do is determine the energy states and wavefunctions for a rotating system consisting of two masses held apart by a rigid bond (cf. Fig. 6-1). We consider the masses to be point particles; the rigid bond holds them at a fixed distance apart in space. We shall not concern ourselves with the translational energy of the system, so we can let the center of mass be fixed at the origin of the coordinate system we use—we are interested only in the rotational energy. We solve the problem by setting up and solving the Schroedinger equation for the system.

To solve this problem it is most convenient if we use spherical polar

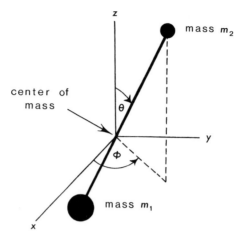

**Fig. 6-1**  Coordinate system for the rigid rotator problem. The two masses $m_1$ and $m_2$ are held by a rigid bond.

coordinates, $r$, $\theta$, $\phi$, defined in Fig. 6-2. The relationship between cartesian and spherical polar coordinates is easily seen by examining the diagram; it is

$$x = r \sin \theta \cos \phi$$
$$y = r \sin \theta \sin \phi \qquad (6\text{-}1)$$
$$z = r \cos \theta$$

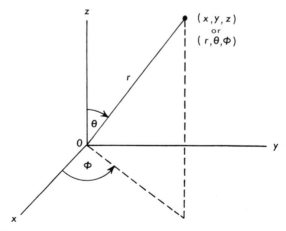

**Fig. 6-2**  Relationship between cartesian coordinates $(x, y, z)$ and spherical polar coordinates $(r, \theta, \phi)$.

If we let the two masses be $m_1$ and $m_2$, we can specify their coordinates as $r_1, \theta, \phi$ and $r_2, \theta, \phi$.

To set up the Schroedinger equation for the system, we need to know the form of the Hamiltonian. This requires that we find expressions for the kinetic and potential energies that can be converted to quantum mechanical operator form. In cartesian coordinates the kinetic energy $T_1$ of mass $m_1$ is

$$T_1 = \frac{m_1}{2} (\dot{x}_1{}^2 + \dot{y}_1{}^2 + \dot{z}_1{}^2) \qquad (6\text{-}2)$$

We have said that we would use spherical polar coordinates in this problem. To convert Eq. (6-2) to spherical polar coordinate form we use Eq. (6-1). The mathematics is straightforward but rather tedious, so we shall not give the details — the result is

$$T_1 = \frac{m_1 r_1{}^2}{2} (\dot{\theta}^2 + \sin^2 \theta \dot{\phi}^2) \qquad (6\text{-}3)$$

Similarly the kinetic energy $T_2$ of mass $m_2$ is

$$T_2 = \frac{m_2 r_2{}^2}{2} (\dot{\theta}^2 + \sin^2 \theta \dot{\phi}^2) \qquad (6\text{-}4)$$

The total kinetic energy $T$ is therefore

$$T = T_1 + T_2 = \frac{m_1 r_1{}^2 + m_2 r_2{}^2}{2} (\dot{\theta}^2 + \sin^2 \theta \dot{\phi}^2) \qquad (6\text{-}5)$$

Now, the so-called moment of inertia $I$ of the system is given by

$$I = m_1 r_1{}^2 + m_2 r_2{}^2 \qquad (6\text{-}6)$$

Therefore,

$$T = \frac{I}{2} (\dot{\theta}^2 + \sin^2 \theta \dot{\phi}^2) \qquad (6\text{-}7)$$

This is obviously the kinetic energy of a single particle of mass $I$ moving on the surface of a sphere of unit radius. Our problem is therefore equivalent to determining the energy states and wavefunctions for such a system.

Provided there are no external forces acting on the system, the potential energy on the surface of the sphere is zero; i.e., $U$ in the Schroedinger equation is zero. To find the Hamiltonian for the system we therefore have only to transform the classical expression for the kinetic energy, Eq. (6-7), to quantum mechanical operator form. If we refer back to Section

4-2, we see that this requires that we first express Eq. (6-7) in terms of the appropriate momenta, $p_\theta$ and $p_\phi$. This is easily done by making use of the definition of momentum in terms of the *Lagrangian function*, which is

$$p_i = \frac{\partial L}{\partial \dot{q}_i} \tag{6-8}$$

Here $L$ is the Lagrangian function which is defined for a conservative system by

$$L(\dot{q}, q) = T(\dot{q}) - U(q) \tag{6-9}$$

$p_i$ are generalized momenta, $q_i$ are generalized coordinates, and $\dot{q}_i$ are generalized velocities, i.e., Eq. (6-9) is valid for any coordinate system. Therefore,

$$p_\theta = \frac{\partial L}{\partial \dot{\theta}} = \frac{\partial T}{\partial \dot{\theta}} = I\dot{\theta} \tag{6-10}$$

$$p_\phi = \frac{\partial L}{\partial \dot{\phi}} = \frac{\partial T}{\partial \dot{\phi}} = I \sin^2 \theta \dot{\phi} \tag{6-11}$$

We now solve Eqs. (6-10) and (6-11) for $\dot{\theta}$ and $\dot{\phi}$ and substitute into Eq. (6-7); we get

$$T = \frac{1}{2I} \left( p_\theta^2 + \frac{p_\phi^2}{\sin^2 \theta} \right) \tag{6-12}$$

This is now in a form from which we could transform it into a quantum mechanical operator by replacing $p_\theta$ by $\hbar/i \ \partial/\partial\theta$ and $p_\phi$ by $\hbar/i \ \partial/\partial\phi$. However, if we did so, we would obtain a non-Hermitian operator. The difficulty is easily overcome by multiplying the first term in the bracket by $\sin \theta / \sin \theta$ before making the transformation; doing this, we get

$$T = \frac{1}{2I} \left[ \frac{1}{\sin \theta} p_\theta (\sin \theta) p_\theta + \frac{p_\phi^2}{\sin^2 \theta} \right] \tag{6-13}$$

We can now make the substitutions to get the Hamiltonian for the system. Since $H = T + U$,

$$\hat{H} = -\frac{\hbar^2}{2I} \left[ \frac{1}{\sin \theta} \frac{\partial}{\partial \theta} \left( \sin \theta \frac{\partial}{\partial \theta} \right) + \frac{1}{\sin^2 \theta} \frac{\partial^2}{\partial \phi^2} \right] \tag{6-14}$$

(Remember that $U = 0$ in this problem.) The Schroedinger equation for the rigid rotator problem is therefore

$$-\frac{\hbar^2}{2I} \left[ \frac{1}{\sin \theta} \frac{\partial}{\partial \theta} \left( \sin \theta \frac{\partial}{\partial \theta} \right) + \frac{1}{\sin^2 \theta} \frac{\partial^2}{\partial \phi^2} \right] \Psi(\theta, \phi) = E\Psi(\theta, \phi) \tag{6-15}$$

Our problem then is to solve Eq. (6-15). This we do by first separating the variables (cf. Section 3-2). We write

$$\Psi(\theta, \phi) = T(\theta)F(\phi) \tag{6-16}$$

Substitute this into Eq. (6-15); after expanding and rearranging the equation, we get

$$F(\phi)\frac{d^2T}{d\theta^2} + F(\phi)\frac{\cos\theta}{\sin\theta}\frac{dT}{d\theta} + T(\theta)\frac{1}{\sin^2\theta}\frac{d^2F}{d\phi^2} + \frac{2IE}{\hbar^2}\Psi(\theta,\phi) = 0 \tag{6-17}$$

Now divide by $\Psi(\theta, \phi)$ to get

$$\frac{1}{T(\theta)}\left(\frac{d^2}{d\theta^2} + \frac{\cos\theta}{\sin\theta}\frac{d}{d\theta}\right)T(\theta) + \frac{1}{F(\phi)\sin^2\theta}\frac{d^2F}{d\phi^2} + \frac{2IE}{\hbar^2} = 0 \tag{6-18}$$

This separates into two ordinary differential equations:

$$\frac{d^2F}{d\phi^2} = -m^2F \tag{6-19}$$

$$\frac{d^2T}{d\theta^2} + \frac{\cos\theta}{\sin\theta}\frac{dT}{d\theta} + \left(\frac{2IE}{\hbar^2} - \frac{m^2}{\sin^2\theta}\right)T = 0 \tag{6-20}$$

$m$ is an arbitrary constant—we have used it in the form $m^2$ for convenience in expressing the final result.

Let us first consider the solution of Eq. (6-19); it is

$$F = e^{im\phi} \times \text{constant} \tag{6-21}$$

Now, although all functions of the form Eq. (6-21) are solutions of Eq. (6-19), only certain ones are allowed in our problem. This is because the functions have to be single-valued, i.e.,

$$e^{im\phi} = e^{im(\phi+2\pi)} \tag{6-22}$$

since adding $2\pi$ to $\phi$ brings us back to the same point in space. Therefore,

$$e^{2\pi mi} = 1 \tag{6-23}$$

Equation (6-23) is true if $m$ is allowed to take only the values

$$m = 0, \pm 1, \pm 2, \pm 3, \ldots \tag{6-24}$$

We now consider that $m$ takes these values in Eq. (6-20) and attempt to solve this second equation. However, before considering the solution, we shall introduce a new variable:

$$s = \cos\theta \tag{6-25}$$

Our reason for changing the variable is that it brings Eq. (6-20) into a standard mathematical form. From Eq. (6-25) we get

$$\frac{dT}{d\theta} = \frac{ds}{d\theta}\frac{dT}{ds} = -\sin\theta\frac{dT}{ds} \tag{6-26}$$

$$\frac{d^2T}{d\theta^2} = \frac{d}{d\theta}\left(\frac{dT}{d\theta}\right) = \sin^2\theta\frac{d^2T}{ds^2} - \cos\theta\frac{dT}{ds} \tag{6-27}$$

Substitute Eqs. (6-26) and (6-27) into Eq. (6-20); we get

$$(1-s^2)\frac{d^2T}{ds^2} - 2s\frac{dT}{ds} + \left(\frac{2IE}{\hbar^2} - \frac{m^2}{1-s^2}\right)T = 0 \tag{6-28}$$

Equation (6-28) is now in a standard mathematical form. It can be solved by a very useful method for solving certain kinds of differential equations. The method involves trying to express the solution in the form of a series; we shall digress to explain it.

## Series Solutions of Differential Equations

There is a certain type of differential equation that occurs very commonly in physical problems – it has the form:

$$\frac{d^2y}{dx^2} + f(x)\frac{dy}{dx} + g(x)y = 0 \tag{6-29}$$

If $f(x)$ and $g(x)$ have certain special forms, these equations are very easy to solve – no doubt you are familiar with the methods for solving simple special equations of this type. However, in general the simpler methods do not work, so we have to resort to trying to find solutions in the form of power series. Certain equations for which we can express the solutions in series form are very important in quantum mechanics – we consider some of them in this chapter. Before we do this, however, we should look at the details of finding series solutions for differential equations. We illustrate the method with a specific example; the example we choose is one that has direct relevance to the problem at hand, i.e., solution of the Schroedinger equation for the rigid rotator.

Consider the equation

$$(1-x^2)\frac{d^2y}{dx^2} - 2x\frac{dy}{dx} + l(l+1)y = 0 \tag{6-30}$$

$l$ is a constant. This equation has the form of Eq. (6-29); it is known as *Legendre's equation*.

We shall try to find a solution of Eq. (6-30) that can be expressed in the form of a series; we write the series

$$y = c_0 + c_1x + c_2x^2 + c_3x^3 + \cdots + c_rx^r + \cdots \tag{6-31}$$

If we can now find the values for the constants $c_r$, we will have solved the problem. Substitute Eq. (6-31) into Eq. (6-30); we get

$$(1-x^2)(2c_2 + 6c_3x + 12c_4x^2 + 20c_5x^3 + \cdots + r(r-1)c_rx^{r-2} + \cdots)$$

$$-2x(c_1 + 2c_2x + 3c_3x^2 + 4c_4x^3 + \cdots + rc_rx^{r-1} + \cdots)$$

$$+ l(l+1)(c_0 + c_1x + c_2x^2 + c_3x^3 + \cdots + c_rx^r + \cdots) = 0 \qquad (6\text{-}32)$$

Equation (6-32) must be true for every value of $x$. This can be so only if the coefficient of each power of $x$ is zero. So let us put these coefficients equal to zero. The constant term gives

$$2c_2 + l(l+1)c_0 = 0$$

i.e.,

$$c_2 = -\frac{l(l+1)}{2}c_0 \qquad (6\text{-}33)$$

The coefficient of $x$ gives

$$6c_3 + (l^2 + l - 2)c_1 = 0$$

i.e.,

$$c_3 = -\frac{(l-1)(l+2)}{6}c_1 \qquad (6\text{-}34)$$

and so on. In general, the coefficient of $x^r$ gives, after some manipulation,

$$c_{r+2} = -\frac{(l-r)(l+r+1)}{(r+2)(r+1)}c_r \qquad (6\text{-}35)$$

What does Eq. (6-35) tell us? It gives us a method for finding $c_{r+2}$ if we know $c_r$. That is, if we know $c_0$, we can find $c_2$, $c_4$, $c_6$, etc. If we know $c_1$, we can find $c_3$, $c_5$, $c_7$, etc. Now, the equation we are trying to solve, Eq. (6-30), is a second-order equation; therefore, it will have two arbitrary constants in its solution. Let us put $c_0$ and $c_1$ as these constants. Using Eqs. (6-31) and (6-35), we can write the general solution for Eq. (6-30) in terms of $c_0$ and $c_1$; it is

$$y = c_0\left[1 - \frac{l(l+1)}{2!}x^2 + \frac{l(l+1)(l-2)(l+3)}{4!}x^4 - \cdots\right]$$
$$+ c_1\left[x - \frac{(l-1)(l+2)}{3!}x^3 + \frac{(l-1)(l+2)(l-3)(l+4)}{5!}x^5 - \cdots\right] \qquad (6\text{-}36)$$

Equation (6-36) is the solution we are looking for. But we have not yet finished since this solution is of interest to us only if the series converge. In general, we would have to examine these series for all values of $x$ and $l$. However, in problems of quantum mechanical interest there are some rather severe restrictions that limit the extent to which we have to examine the convergence properties. Usually, when equations of type Eq. (6-30) appear in physical problems the variable $x$ is the cosine of an angle—the range of $x$ is therefore $-1 \leqslant x \leqslant 1$. Since wavefunctions must be finite, we are interested in solutions that remain finite in this range. We shall not examine the solutions to determine the acceptable ones; if we did, we would discover that we have to limit $l$ to non-negative, integral values. Under this condition, one or other of the series in Eq. (6-36) becomes a polynomial, whilst the other diverges at $x = \pm 1$. We can therefore write a set of solutions for Eq. (6-30) that will be acceptable in physical problems—there will be a solution for each non-negative integral value of $l$. The solutions will contain the arbitrary constants $c_0$ and $c_1$—if we choose these so that $y = 1$ when $x = 1$, we get what are known as *Legendre polynomials*. The Legendre polynomial of degree $l$ is written $P_l(x)$. By using what is known as *Rodrigues' formula*, we can write the Legendre polynomials very concisely†:

$$P_l(x) = \frac{1}{2^l l!}\frac{d^l}{dx^l}(x^2 - 1)^l \qquad (6\text{-}37)$$

†For proof that Eq. (6-37) is indeed a correct expression for the Legendre polynomials see, for example, reference E1 in the Bibliography, p. 543.

We shall now return to consideration of the equation we wanted to solve, Eq. (6-28). We shall write it down again:

$$(1-s^2)\frac{d^2T}{ds^2} - 2s\frac{dT}{ds} + \left(\frac{2IE}{\hbar^2} - \frac{m^2}{1-s^2}\right)T = 0$$

We could solve the equation by applying the method of series solution directly. However, we can use the example we have just finished examining and save ourselves some work. Consider again the Legendre equation, Eq. (6-30):

$$(1-x^2)\frac{d^2y}{dx^2} - 2x\frac{dy}{dx} + l(l+1)y = 0$$

Differentiate it $m$ times; we get

$$(1-x^2)\frac{d^{m+2}y}{dx^{m+2}} - 2x(m+1)\frac{d^{m+1}y}{dx^{m+1}} + [l(l+1) - m(m+1)]\frac{d^my}{dx^m} = 0 \qquad (6\text{-}38)$$

Let us now introduce a new function:

$$q = \frac{d^my}{dx^m} \qquad (6\text{-}39)$$

Equation (6-38) becomes

$$(1-x^2)\frac{d^2q}{dx^2} - 2x(m+1)\frac{dq}{dx} + [l(l+1) - m(m+1)]q = 0 \qquad (6\text{-}40)$$

For reasons that will soon be apparent, we now put

$$q = (1-x^2)^{-m/2}p \qquad (6\text{-}41)$$

where, from Eq. (6-39),

$$p = (1-x^2)^{m/2}\frac{d^my}{dx^m} \qquad (6\text{-}42)$$

Substitute Eq. (6-41) into Eq. (6-40); we get, after some manipulation,

$$(1-x^2)\frac{d^2p}{dx^2} - 2x\frac{dp}{dx} + \left[l(l+1) - \frac{m^2}{1-x^2}\right]p = 0 \qquad (6\text{-}43)$$

This is the equation we were trying to obtain, because it has the same form as the one we are trying to solve. The solution obviously is the function $p$ which is given by Eq. (6-42). But $y$ in Eq. (6-42) is just the solution of the

Legendre equation, and we recall that the solution of this latter equation that interested us was the Legendre polynomial $P_l(x)$. Therefore,

$$p = (1-x^2)^{m/2} \frac{d^m}{dx^m} P_l(x) \qquad (6\text{-}44)$$

This function is called an *associated Legendre polynomial of degree l and order m*—we give it the symbol $P_l^m(x)$.

We can conclude then that a solution to the equation we are trying to solve, Eq. (6-28), is $P_l^m(s)$. Remembering that $s = \cos\theta$ (cf. Eq. (6-25)), this solution is, therefore,

$$T = P_l^{|m|}(\cos\theta) \qquad (6\text{-}45)$$

Here we have written $|m|$ because $m$ can take positive or negative values (cf. Eq. (6-24)).

To determine the possible energies for the system we just have to compare Eqs. (6-28) and (6-43); this gives us

$$\frac{2IE}{\hbar^2} = l(l+1)$$

or

$$E_l = \frac{l(l+1)\hbar^2}{2I} \qquad (6\text{-}46)$$

We get the eigenfunctions corresponding to these eigenvalues by combining Eqs. (6-21) and (6-45); they are

$$\Psi_{l,m}(\theta,\phi) = NP_l^{|m|}(\cos\theta)e^{im\phi} \qquad (6\text{-}47)$$

Here $N$ is a normalization constant, which we shall discuss in a moment. The functions $\Psi_{l,m}$ are well-known to mathematicians—they are called *spherical harmonics* and are usually given the symbol $Y_{l,m}$. We should recall that we met these functions before when we dealt with angular momentum (cf. Section 5-3).

The functions Eq. (6-47) must be physically acceptable. This places some restrictions on the values of $l$ and $m$. We shall not examine the details of these restrictions, but we can note that $P_l^m(\cos\theta)$ remains finite in the range that $\cos\theta$ can cover if $l$ is a non-negative integer and $l \geq |m|$. We have already determined that $m$ can take only integral values; therefore,

$$m = -l, -l+1, \ldots, -1, 0, 1, \ldots, l \qquad (6\text{-}48)$$

## 6-3  ORTHOGONALITY AND NORMALIZATION OF THE SOLUTIONS OF THE RIGID ROTATOR PROBLEM

In our earlier discussions on the wavefunctions for quantum mechanical systems we saw that these form orthonormal sets of functions (cf. Section 4-7). Let us briefly examine the orthonormality properties of the solutions of the rigid rotator problem. We shall not consider all the details, but merely present sufficient to give some idea of the mathematical manipulations needed to deal with our solution functions. We look first at the orthogonality properties.

If the solution functions are orthogonal in the present case,

$$\int_0^{2\pi} \int_0^{\pi} Y^*_{l',m'} Y_{l,m} \sin\theta d\theta d\phi = 0 \tag{6-49}$$

unless $l$ and $m$ are equal to $l'$ and $m'$. What in fact is the value of the integral in Eq. (6-49)? If we substitute for the spherical harmonics from Eq. (6-47), we have to show that the following integral is zero:

$$\int_{-1}^{1} P_{l'}^{|m'|}(x)\, P_l^{|m|}(x)\, dx \int_0^{2\pi} e^{i(m-m')\phi}\, d\phi \tag{6-50}$$

Here we have substituted $x = \cos\theta$, in which case $dx = -\sin\theta\, d\theta$.

Now the integral over $\phi$ in Eq. (6-50) is zero if $m \neq m'$. Why is this?

$$\int_0^{2\pi} e^{i(m-m')\phi} d\phi = \frac{e^{i(m-m')\phi}}{i(m-m')} \Bigg|_0^{2\pi} \tag{6-51}$$

Since $m$ and $m'$ are integers,

$$e^{i(m-m')2\pi} = 1 \qquad \text{if } m \neq m' \tag{6-52}$$

and the whole integral in Eq. (6-50) is therefore zero because of the zero value of Eq. (6-51).

If $m = m'$, we can show that the integral in Eq. (6-50) is still zero unless $l = l'$. We do this as follows:

$$\int_{-1}^{1} P_{l'}^{|m'|}(x) P_l^{|m|}(x) dx \int_0^{2\pi} e^{i(m-m')\phi} d\phi = 2\pi \int_{-1}^{1} P_{l'}^{|m|}(x) P_l^{|m|}(x)\, dx \tag{6-53}$$

since

$$\int_0^{2\pi} e^{i(m-m')\phi}\, d\phi = 2\pi \qquad \text{if } m = m' \tag{6-54}$$

Therefore, we have to show that the right side of Eq. (6-53) equals zero. To do this we start from Eq. (6-43), the defining equation for the associated Legendre polynomials; this equation can be written

$$(1-x^2)\frac{d^2 P_l{}^m}{dx^2} - 2x\frac{dP_l{}^m}{dx} + \left[l(l+1) - \frac{m^2}{1-x^2}\right]P_l{}^m = 0 \qquad (6\text{-}55)$$

Multiply Eq. (6-55) by $P_{l'}{}^m$; we get

$$(1-x^2)P_{l'}{}^m\frac{d^2 P_l{}^m}{dx^2} - 2xP_{l'}{}^m\frac{dP_l{}^m}{dx} + \left[l(l+1) - \frac{m^2}{1-x^2}\right]P_l{}^m P_{l'}{}^m = 0 \qquad (6\text{-}56a)$$

Similarly,

$$(1-x^2)P_l{}^m\frac{d^2 P_{l'}{}^m}{dx^2} - 2xP_l{}^m\frac{dP_{l'}{}^m}{dx} + \left[l'(l'+1) - \frac{m^2}{1-x^2}\right]P_l{}^m P_{l'}{}^m = 0 \quad (6\text{-}56b)$$

Subtract Eq. (6-56b) from Eq. (6-56a) and integrate between 1 and $-1$; we get

$$\int_{-1}^{1} \left\{(1-x^2)\left[P_{l'}{}^m\frac{d^2 P_l{}^m}{dx^2} - P_l{}^m\frac{d^2 P_{l'}{}^m}{dx^2}\right] - 2x\left[P_{l'}{}^m\frac{dP_l{}^m}{dx} - P_l{}^m\frac{dP_{l'}{}^m}{dx}\right]\right\} dx$$

$$+ \left[l(l+1) - l'(l'+1)\right]\int_{-1}^{1} P_l{}^m P_{l'}{}^m \, dx = 0 \qquad (6\text{-}57)$$

i.e.,

$$\int_{-1}^{1} \frac{d}{dx}\left\{(1-x^2)\left[P_{l'}{}^m\frac{dP_l{}^m}{dx} - P_l{}^m\frac{dP_{l'}{}^m}{dx}\right]\right\} dx$$

$$+ \left[l(l+1) - l'(l'+1)\right]\int_{-1}^{1} P_l{}^m P_{l'}{}^m \, dx = 0 \qquad (6\text{-}58)$$

The first integral is zero. Therefore, provided $l \neq l'$ (which was our original condition)

$$\int_{-1}^{1} P_l{}^m(x)P_{l'}{}^m(x)dx = 0 \qquad (6\text{-}59)$$

i.e., the right side of Eq. (6-53) is zero, and we have verified the orthogonality property.

In fact, when $l = l'$ and $m = m'$ we have the normalization condition

$$\int Y_{l,m}^* Y_{l,m} d\tau = 1 \qquad (6\text{-}60)$$

Substitute again from Eq. (6-47); we get, if we refer to Eq. (6-53) and put back in the normalization constant,

$$N^* N 2\pi \int_{-1}^{1} [P_l{}^{|m|}(x)]^2 \, dx = 1$$

We shall not work out this integral since the mathematics is somewhat tedious and no new quantum mechanical ideas are involved. If we did work it out, we would determine the normalization constant to be

$$N = \left[ \frac{(2l+1)}{4\pi} \frac{(l-|m|)!}{(l+|m|)!} \right]^{1/2} \tag{6-61}$$

We can now at last write the complete solution for the rigid rotator equation; it is

$$\Psi_{l,m}(\theta, \phi) = Y_{l,m}(\theta, \phi)$$

$$= \left[ \frac{(2l+1)}{4\pi} \frac{(l-|m|)!}{(l+|m|)!} \right]^{1/2} P_l^{|m|}(\cos \theta) e^{im\phi} \tag{6-62}$$

## 6-4  THE HARMONIC OSCILLATOR

The problem of the harmonic oscillator is another problem of interest to chemists: atoms bonded together in molecules have vibrational motions that are approximately similar to that which we are about to discuss. In fact, if the amplitude of vibration of a diatomic molecule is not too large, the harmonic oscillator is sometimes an acceptable approximation.

To determine the eigenstates of the system, we again have to solve a Schroedinger equation. This means that we have to determine the form of the Hamiltonian for the system: we first determine the form of the potential energy part of the Hamiltonian; then we combine this with the kinetic energy term and fit the total Hamiltonian into the Schroedinger equation. Solution of the equation gives us the possible wavefunctions and corresponding energies for the system.

A convenient system to use in the treatment of the harmonic oscillator problem is a vibrating particle of mass $m$ attached to a linear spring as shown in Fig. 6-3. (In fact, we could as a first approximation use as a model for the diatomic molecule two atoms held together by a spring.) We say that the vibrational motion of the particle is *simple harmonic motion* if the restoring force $f$ acting on the particle is directly proportional to the displacement $x$ from equilibrium, i.e.,

$$f = -kx \tag{6-63}$$

Here $k$ is a constant called the *force constant* — it is a measure of the stiffness of the spring (or bond, in molecular problems). Note that this is a one-dimensional problem.

Now, if the system is a conservative system, we can easily show that if

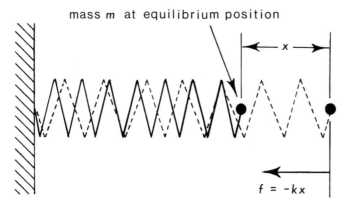

mass *m* at equilibrium position

$f = -kx$

**Fig. 6-3**  A harmonic oscillator. The mass *m* is attached to the linear spring which exerts a restoring force *f* when the mass is displaced from its equilibrium position.

$U(x)$ is the potential energy function,

$$f = -\frac{dU(x)}{dx} \qquad (6\text{-}64)$$

According to Newton's law,

$$f = m\ddot{x} \qquad (6\text{-}65)$$

Therefore, if Eq. (6-64) is true,

$$m\frac{d\dot{x}}{dt} = -\frac{dU(x)}{dx} \qquad (6\text{-}66)$$

Integrating, we get

$$m\int \frac{d\dot{x}}{dt}\,dx = -\int \frac{dU(x)}{dx}\,dx \qquad (6\text{-}67)$$

This gives

$$\frac{m\dot{x}^2}{2} + U(x) = c \qquad (6\text{-}68)$$

where $c$ is the constant of integration. We have therefore proved that the sum of the kinetic and potential energies is constant, independent of time – this is just the familiar definition of a conservative system, so we can accept Eq. (6-64) since it is consistent with this definition.

Equating Eqs. (6-63) and (6-64), we get

$$\frac{dU}{dx} = kx \qquad (6\text{-}69)$$

i.e.,

$$U = \frac{kx^2}{2} \tag{6-70}$$

Equation (6-70) is a parabola. In Fig. 6-4 we show a plot of a function of this type.

As we have mentioned before, the total energy of the system is given by Hamilton's function $H$, the sum of the kinetic energy $T$ and the potential energy $U$:

$$H = T + U$$

Therefore, using Eq. (6-70), we get

$$H = \frac{p_x^2}{2m} + \frac{kx^2}{2} \tag{6-71}$$

$p_x$ is the momentum of the particle; we have used the subscript to emphasize that we are treating a one-dimensional problem. To convert Eq. (6-71) to the quantum mechanical operator we apply the rules for doing this (cf. Section 4-2); we get

$$\hat{H} = -\frac{\hbar^2}{2m}\frac{d^2}{dx^2} + \frac{kx^2}{2} \tag{6-72}$$

We can use ordinary differentials here since we are working with just one variable, $x$.

Using Eq. (6-72), we can write the Schroedinger equation for the problem; it is

$$\frac{d^2\Psi}{dx^2} + \frac{2m}{\hbar^2}\left(E - \frac{kx^2}{2}\right)\Psi = 0 \tag{6-73}$$

To save our constantly writing $2mE/\hbar^2$ we replace this by $\alpha$. Equation (6-73) then becomes

$$\frac{d^2\Psi}{dx^2} + \left(\alpha - \frac{km}{\hbar^2}x^2\right)\Psi = 0 \tag{6-74}$$

Next, we change from the variable $x$ to a new variable by means of the transformation

$$y = \sqrt{\beta}x \quad \text{where} \quad \beta = \frac{\sqrt{km}}{\hbar} \tag{6-75}$$

This gives us

$$\frac{d^2\Psi}{dx^2} = \frac{d^2\Psi}{dy^2}\beta \tag{6-76}$$

Substitute Eqs. (6-75) and (6-76) into Eq. (6-74); we get

$$\frac{d^2\Psi}{dy^2} + \left(\frac{\alpha}{\beta} - y^2\right)\Psi = 0 \tag{6-77}$$

We can solve Eq. (6-77) by making an appropriate substitution. We determine the substitution by considering the functions

$$\Psi = \exp\left(\pm\frac{y^2}{2}\right) \tag{6-78}$$

These are solutions of the equations

$$\frac{d^2\Psi}{dy^2} + (\mp 1 - y^2)\Psi = 0 \tag{6-79}$$

Now, when $y^2$ becomes large the $\mp 1$ in Eq. (6-79) and the term $\alpha/\beta$ in Eq. (6-77) are insignificant—the equations are approximately the same under this condition. We therefore try to solve Eq. (6-77) by substituting a function of the form

$$\Psi = u(y) \exp\left(-\frac{y^2}{2}\right) \tag{6-80}$$

(We have rejected the function $\exp(+y^2/2)$ because this tends to infinity as $y$—and consequently $x$, the displacement from equilibrium—tends to infinity.) Differentiating Eq. (6-80), we get

$$\frac{d^2\Psi}{dy^2} = \left(\frac{d^2u}{dy^2} - 2y\frac{du}{dy} + y^2u - u\right)\exp\left(-\frac{y^2}{2}\right) \tag{6-81}$$

Substitute Eqs. (6-81) and (6-80) into Eq. (6-77). If we then write $2v$ for $\alpha/\beta - 1$, we get

$$\frac{d^2u}{dy^2} - 2y\frac{du}{dy} + 2vu = 0 \tag{6-82}$$

We can now understand why we have made the substitutions and changed the variable: Eq. (6-82) is a standard mathematical equation; it is called *Hermite's equation.*

We can solve Eq. (6-82) by the series method we outlined in Section 6-2. However, we shall not go through this solution since the ideas involved are similar to those we used to solve the rigid rotator problem. Equation (6-82) has a general solution that can be expressed as a superposition of two infinite series. The physical requirements of the form of the wavefunctions, however, limit the solutions to the functions we get when $v$ is an integer. In this case the series that are the solutions of Hermite's equation give only one acceptable form for the solution of our problem; this is

$$u = H_v(y) = (2y)^v - \frac{v(v-1)}{1!}(2y)^{v-2}$$

$$+ \frac{v(v-1)(v-2)(v-3)}{2!}(2y)^{v-4} + \cdots \qquad (6\text{-}83)$$

This series is continued as far as the constant term, or as far as the lowest positive power of $y$, whichever is applicable; i.e., Eq. (6-83) is a polynomial — we call it a *Hermite polynomial of degree v* and label it $H_v(y)$. We can write the general definition of Hermite polynomials very concisely as follows:

$$H_v(y) = (-1)^v \exp(y^2) \frac{d^v}{dy^v}[\exp(-y^2)] \qquad (6\text{-}84)$$

If we now substitute the Hermite polynomials into Eq. (6-80), we get a solution function for the harmonic oscillator problem for each possible value of $v$. Making the substitution, we get

$$\Psi_v(y) = N_v H_v(y) \exp\left(-\frac{y^2}{2}\right) \qquad (6\text{-}85)$$

Here we have introduced a normalizing constant $N_v$. To determine the form for $N_v$ we would have to consider the normalization condition

$$N_v^* N_v \int_{-\infty}^{+\infty} [H_v(y)]^2 \exp(-y^2)dy = 1 \qquad (6\text{-}86)$$

We shall not go into the details of working out this integral; the result, if we did work it out would be

$$N_v = \left(\frac{1}{\pi}\right)^{1/4}(2^v v!)^{-1/2} \qquad (6\text{-}87)$$

Therefore, we get for the eigenfunctions of the harmonic oscillator

$$\Psi_v(y) = \left(\frac{1}{\pi}\right)^{1/4}(2^v v!)^{-1/2} H_v(y) \exp\left(-\frac{y^2}{2}\right) \qquad (6\text{-}88)$$

where

$$y = \sqrt{\beta}x \quad \text{and} \quad \beta = \frac{\sqrt{km}}{\hbar}$$

The form of the Hermite polynomials can easily be determined by substituting values for $v$ into Eq. (6-83). If we did this, we would find that the first few values are:

$$H_0(y) = 1$$

$$H_1(y) = 2y$$

$$H_2(y) = 4y^2 - 2 \tag{6-89}$$

$$H_3(y) = 8y^3 - 12y$$

$$H_4(y) = 16y^4 - 48y^2 + 12$$

Figure 6-4 shows what the eigenfunctions of the harmonic oscillator look like when plotted.

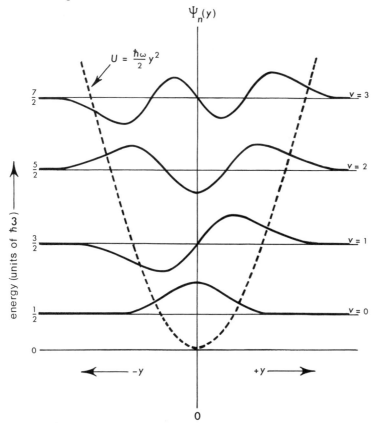

**Fig. 6-4**   Eigenfunctions and energies for the first few states of the harmonic oscillator. The eigenfunctions are superimposed on the corresponding energy levels. Also shown, is the potential energy function $U$ (dashed curve).

To complete our consideration of the harmonic oscillator problem we shall determine the energies of the states described by Eq. (6-88). This we can do by recalling that $\alpha/\beta - 1 = 2v$ ($v$ is a positive integer or zero). Substituting for $\alpha$ and $\beta$ and rearranging, we get

$$E_v = \left(v + \frac{1}{2}\right)\hbar\left(\frac{k}{m}\right)^{1/2} \tag{6-90}$$

We can express this in terms of the angular frequency of the oscillator. You should recall from your elementary physics courses that this latter quantity is given by

$$\omega = 2\pi v = \left(\frac{k}{m}\right)^{1/2} \tag{6-91}$$

$v$ is the frequency of vibration. Now substitute Eq. (6-91) into Eq. (6-90); we get

$$E_v = (v + \tfrac{1}{2})\hbar\omega \qquad \text{where } v = 0, 1, 2, 3, \ldots \tag{6-92}$$

This means that the energy is quantized since $v$ can take only the values shown. Also, the zero-point energy is given by Eq. (6-92) with $v = 0$, i.e.,

$$E_0 = \tfrac{1}{2}\hbar\omega \tag{6-93}$$

## 6-5   THE HYDROGEN ATOM

The last example of a solution of a Schroedinger equation that we shall consider here is the hydrogen atom problem. The hydrogen atom, of course, is the simplest chemical system. Yet even for this the quantum mechanical problem involves considerable mathematical manipulation.

As for the previous examples, what we shall do is set up the Schroedinger equation for the system; then we shall solve this to determine the form of the wavefunctions and expressions for the energies associated with these functions.

The system we shall be dealing with is shown in Fig. 6-5. This diagram needs some explanation. First, you will note that we have generalized the problem a little by using a nucleus of charge $Ze$ — in this way we can solve the problem for the hydrogen-like ions ($He^+$, $Li^{2+}$, etc.) as well as for hydrogen itself. For hydrogen, of course, $Z = 1$. The mass of the nucleus is $M$. The electron's mass and charge have been given the conventional symbols $m$ and $-e$. The distance between the nucleus and the electron is $r$.

You will note from Fig. 6-5 that we have placed the origin for the coordinate systems at the nucleus. To be strictly correct the origin should be

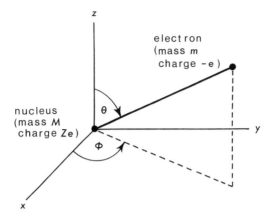

**Fig. 6-5**   Coordinate system for the hydrogen-like atoms. The nucleus is at the origin and is assumed to be at the center of mass.

at the center of mass. However, the nucleus is very much larger than the electron, so no great error is introduced by considering the center of mass to be at the nucleus. Furthermore, the reduced mass $\mu$ of the system is almost equal to the electron's mass. (For the hydrogen atom, $\mu = mM/m + M = (1846/1847)m$.) We can therefore very closely approximate the system by one in which an electron moves in a spherically symmetric field created by the nucleus at the origin of the coordinate system. Also, we can consider the nucleus to be at rest and concern ourselves only with the internal motion of the system — this is because the potential energy is a function only of the internal coordinates of the system.

To determine the Hamiltonian we need to know the form of the potential energy function. The potential energy results from the attraction between the nucleus and the electron, i.e.,

$$U = -\frac{Ze^2}{r} \tag{6-94}$$

Therefore, the Schroedinger equation is

$$\nabla^2\Psi + \frac{2m}{\hbar^2}\left(E + \frac{Ze^2}{r}\right)\Psi = 0 \tag{6-95}$$

Since the problem has spherical symmetry, we can simplify it by converting Eq. (6-95) to spherical polar coordinate form. This necessitates conversion of the Laplacian operator to spherical polar coordinate form. The mathematics required to do this is straightforward, but it does involve a considerable amount of manipulation, so we shall just give the result; it is†

$$\nabla^2 = \frac{\partial^2}{\partial x^2} + \frac{\partial^2}{\partial y^2} + \frac{\partial^2}{\partial z^2}$$

$$= \frac{\partial^2}{\partial r^2} + \frac{2}{r}\frac{\partial}{\partial r} + \frac{1}{r^2}\frac{\partial^2}{\partial\theta^2} + \frac{\cos\theta}{r^2\sin\theta}\frac{\partial}{\partial\theta} + \frac{1}{r^2\sin^2\theta}\frac{\partial^2}{\partial\phi^2} \qquad (6\text{-}96)$$

Inserting Eq. (6-96) into Eq. (6-95), we get

$$\left(\frac{\partial^2}{\partial r^2} + \frac{2}{r}\frac{\partial}{\partial r} + \frac{1}{r^2}\frac{\partial^2}{\partial\theta^2} + \frac{\cos\theta}{r^2\sin\theta}\frac{\partial}{\partial\theta} + \frac{1}{r^2\sin^2\theta}\frac{\partial^2}{\partial\phi^2}\right)\Psi + \frac{2m}{\hbar^2}\left(E + \frac{Ze^2}{r}\right)\Psi = 0$$
$$(6\text{-}97)$$

We can solve Eq. (6-97) by first partially separating the variables. We write

$$\Psi = R(r)P(\theta,\phi) \qquad (6\text{-}98)$$

Substitute this into Eq. (6-97) and divide by $\Psi$, i.e., follow the standard procedure we outlined in Section 3-2; we get

$$\frac{1}{R}\left(r^2\frac{d^2R}{dr^2} + 2r\frac{dR}{dr}\right) + \frac{2mr^2}{\hbar^2}\left(E + \frac{Ze^2}{r}\right) + \frac{1}{P\sin\theta}\left(\sin\theta\frac{\partial^2P}{\partial\theta^2} + \cos\theta\frac{\partial P}{\partial\theta}\right)$$

$$+ \frac{1}{P\sin^2\theta}\frac{\partial^2P}{\partial\phi^2} = 0 \qquad (6\text{-}99)$$

Continuing with the separation of variables technique, we put the terms containing $\theta$ and $\phi$ equal to a constant; in anticipation of the result, we call this constant $-l(l+1)$. This gives us

$$\left(\frac{\partial^2}{\partial\theta^2} + \frac{\cos\theta}{\sin\theta}\frac{\partial}{\partial\theta} + \frac{1}{\sin^2\theta}\frac{\partial^2}{\partial\phi^2}\right)P(\theta,\phi) + l(l+1)P(\theta,\phi) = 0 \qquad (6\text{-}100)$$

We can write this equation in a slightly different form:

$$\left[\frac{1}{\sin\theta}\frac{\partial}{\partial\theta}\left(\sin\theta\frac{\partial}{\partial\theta}\right) + \frac{1}{\sin^2\theta}\frac{\partial^2}{\partial\phi^2}\right]P(\theta,\phi) + l(l+1)P(\theta,\phi) = 0 \qquad (6\text{-}101)$$

The reason for writing the equation this way is so that we can compare it to the equation for the rigid rotator, Eq. (6-15). Obviously, it has the same form. The solutions, therefore, are the spherical harmonics; i.e.,

$$P(\theta,\phi) = Y_{l,m}(\theta,\phi) \qquad (6\text{-}102)$$

---

†For a simple outline of the conversion of the Laplacian operator to spherical polar coordinates you can consult reference D4 in the Bibliography, p. 263.

where

$$l = 0, 1, 2, 3, \ldots$$
$$m = 0, \pm 1, \pm 2, \pm 3, \ldots, \pm l \tag{6-103}$$

Equation (6-98) therefore becomes

$$\Psi = R(r) Y_{l,m}(\theta, \phi) \tag{6-104}$$

We now turn our attention to the form of $R(r)$. This we determine by considering the other part of Eq. (6-99). The terms containing $r$ must equal the negative of the same constant that we have already defined; i.e.,

$$\frac{1}{R}\left(r^2 \frac{d^2 R}{dr^2} + 2r \frac{dR}{dr}\right) + \frac{2mr^2}{\hbar^2}\left(E + \frac{Ze^2}{r}\right) = l(l+1) \tag{6-105}$$

Therefore,

$$\frac{d^2 R}{dr^2} + \frac{2}{r}\frac{dR}{dr} + \left(\frac{2mE}{\hbar^2} + \frac{2mZe^2}{\hbar^2 r} - \frac{l(l+1)}{r^2}\right) R = 0 \tag{6-106}$$

We can solve Eq. (6-106) most easily by introducing a new variable defined by

$$x = \frac{2mZe^2 r}{n\hbar^2} \tag{6-107}$$

$n$ is a new parameter defined by

$$n = \left(-\frac{mZ^2 e^4}{2\hbar^2 E_n}\right)^{1/2} \tag{6-108}$$

This converts Eq. (6-106) to

$$\frac{d^2 R}{dx^2} + \frac{2}{x}\frac{dR}{dx} + \left[\frac{n}{x} - \frac{l(l+1)}{x^2} - \frac{1}{4}\right] R = 0 \tag{6-109}$$

We can solve Eq. (6-109) by making an appropriate guess at the solution— as we did for the harmonic oscillator. The form of the solution is indicated to us if we consider what happens when $x$ becomes very large — Eq. (6-109) reduces to

$$\frac{d^2 R}{dx^2} - \frac{R}{4} = 0 \tag{6-110}$$

The solutions of Eq. (6-110) are $\exp(\pm x/2)$. Now, a solution of the form $\exp(+x/2)$ would lead to eigenfunctions that could not be normalized; therefore, we reject this solution immediately and try a solution of the form

$$R = g(x) \exp\left(-\frac{x}{2}\right) \tag{6-111}$$

When we substitute Eq. (6-111) into Eq. (6-109) we get

$$\frac{d^2g}{dx^2} + \left(\frac{2}{x} - 1\right)\frac{dg}{dx} + \left(\frac{n-1}{x} - \frac{l(l+1)}{x^2}\right)g = 0 \tag{6-112}$$

We can try to solve Eq. (6-112) by the series method of Section 6-2. Again, as in the harmonic oscillator problem, we shall not go through the details of the solution since we have already illustrated the method when dealing with the rigid rotator.

When we try the series expansion method on Eq. (6-112) we end up with another differential equation:

$$x\frac{d^2f}{dx^2} + (2l+2-x)\frac{df}{dx} + (n-l-1)f = 0 \tag{6-113}$$

where

$$g(x) = x^l f(x) \tag{6-114}$$

It seems, therefore, that we are no better off. However, Eq. (6-113) is a standard mathematical form for which the solution is well-known to mathematicians. In fact, this equation, too, is susceptible to the series expansion method. The acceptable solutions (i.e., those giving physically acceptable wavefunctions) are the so-called associated Laguerre polynomials; the *associated Laguerre polynomial of degree* $\alpha - \beta$ is written $L_\alpha^\beta(x)$ where $\beta = 2l+1$ and $\alpha = n+l$. $\alpha$ and $\beta$ are required to be integers: $l$ and $n$ must also be integers. Also, $n$ must be greater than $l$.

The form of $R(r)$, which is the radial part of the wavefunction, is given by substitution into Eqs. (6-111) and (6-114); if we do this, we get

$$R(r) = -Nx^l \exp\left(-\frac{x}{2}\right)L_{n+l}^{2l+1}(x) \tag{6-115}$$

$N$ is a normalization constant. $x$, of course, was not our original variable. To return to $r$, the variable we started with, we go back to the defining equation for $x$, Eq. (6-107). When we do this we can make the resulting form of $R(r)$ more compact by following the convention of using the first Bohr radius $a_0$, which is the radius of the first orbit in the old Bohr theory of the hydrogen atom — this eliminates some constants since $a_0$ is given by

$$a_0 = \frac{\hbar^2}{me^2}$$

Substitution of this into Eq. (6-107) gives

$$x = \frac{2Zr}{na_0} \tag{6-116}$$

The complete wavefunctions for the hydrogen-like atoms include the normalization constant $N$. Again, as for the harmonic oscillator problem, we shall not work this out. However, if we did, we would determine $N$ to be

$$N = \left[\left(\frac{2Z}{na_0}\right)^3 \frac{(n-l-1)!}{2n[(n+l)!]^3}\right]^{1/2} \tag{6-117}$$

We can now write the complete wavefunctions for the hydrogen-like atoms. First, we substitute Eqs. (6-116) and (6-117) into Eq. (6-115); then we substitute the result along with Eq. (6-102) into Eq. (6-98). This gives for the wavefunctions:

$$\Psi_{n,l,m} = -\left[\left(\frac{2Z}{na_0}\right)^3 \frac{(n-l-1)!}{2n[(n+l)!]^3}\right]^{1/2} \left(\frac{2Zr}{na_0}\right)^l \exp\left(-\frac{Zr}{na_0}\right) L_{n+l}^{2l+1}\left(\frac{2Zr}{na_0}\right) Y_{l,m}(\theta,\phi) \tag{6-118}$$

To obtain the wavefunctions for the various specific states of the atoms we just have to put the appropriate values for $n$, $l$, and $m$ into Eq. (6-118). Doing this for the first few states, we get

$$\Psi_{1,0,0} = \frac{1}{\sqrt{\pi}}\left(\frac{Z}{a_0}\right)^{3/2} \exp\left(-\frac{Zr}{a_0}\right)$$

$$\Psi_{2,0,0} = \frac{1}{4\sqrt{2\pi}}\left(\frac{Z}{a_0}\right)^{3/2}\left(2-\frac{Zr}{a_0}\right)\exp\left(-\frac{Zr}{2a_0}\right)$$

$$\Psi_{2,1,0} = \frac{1}{4\sqrt{2\pi}}\left(\frac{Z}{a_0}\right)^{3/2}\frac{Zr}{a_0}\exp\left(-\frac{Zr}{2a_0}\right)\cos\theta$$

$$\Psi_{2,1,\pm1} = \frac{1}{8\sqrt{\pi}}\left(\frac{Z}{a_0}\right)^{3/2}\frac{Zr}{a_0}\exp\left(-\frac{Zr}{2a_0}\right)\sin\theta\exp(\pm i\phi)$$

$$\Psi_{3,0,0} = \frac{1}{81\sqrt{3\pi}}\left(\frac{Z}{a_0}\right)^{3/2}\left(27-18\frac{Zr}{a_0}+2\frac{Z^2r^2}{a_0}\right)\exp\left(-\frac{Zr}{3a_0}\right) \tag{6-119}$$

$$\Psi_{3,1,0} = \frac{\sqrt{2}}{81\sqrt{\pi}}\left(\frac{Z}{a_0}\right)^{3/2}\left(6-\frac{Zr}{a_0}\right)\frac{Zr}{a_0}\exp\left(-\frac{Zr}{3a_0}\right)\cos\theta$$

$$\Psi_{3,1,\pm1} = \frac{1}{81\sqrt{\pi}}\left(\frac{Z}{a_0}\right)^{3/2}\left(6-\frac{Zr}{a_0}\right)\frac{Zr}{a_0}\exp\left(-\frac{Zr}{3a_0}\right)\sin\theta\exp(\pm i\phi)$$

$$\Psi_{3,2,0} = \frac{1}{81\sqrt{6\pi}}\left(\frac{Z}{a_0}\right)^{3/2}\frac{Z^2r^2}{a_0^2}\exp\left(-\frac{Zr}{3a_0}\right)(3\cos^2\theta-1)$$

$$\Psi_{3,2,\pm1} = \frac{1}{81\sqrt{\pi}}\left(\frac{Z}{a_0}\right)^{3/2}\frac{Z^2r^2}{a_0^2}\exp\left(-\frac{Zr}{3a_0}\right)\sin\theta\cos\theta\exp(\pm i\phi)$$

$$\Psi_{3,2,\pm2} = \frac{1}{162\sqrt{\pi}}\left(\frac{Z}{a_0}\right)^{3/2}\frac{Z^2r^2}{a_0^2}\exp\left(-\frac{Zr}{3a_0}\right)\sin^2\theta\exp(\pm2i\phi)$$

How do we determine the energies corresponding to the states for which we have given the wavefunctions in Eq. (6-118)? Look back to the equation defining the parameter $n$, i.e., Eq. (6-108). Rearrangement of this gives

$$E_n = -\frac{me^4 Z^2}{2n^2 \hbar^2} \qquad (6\text{-}120)$$

Remember here that $n$ must be an integer and must be greater than $l$.

We see, therefore, that three quantum numbers appear in the solution to the hydrogen-like atom problem; they are given by

$$n = 1, 2, 3, 4, \ldots$$
$$l = 0, 1, 2, 3, \ldots, (n-1) \qquad (6\text{-}121)$$
$$m = 0, \pm 1, \pm 2, \ldots, \pm l$$

Before we leave this problem we should briefly examine the significance of the results we have obtained. First we see that the energy depends only on the principal quantum number $n$ (cf. Eq. (6-120)). But there are generally several values of $l$ and $m$ that correspond to a single value of $n$. Therefore, in general each state of the system is degenerate. We can determine the extent of the degeneracy by noting that there are $2l + 1$ possible values of $m$ for each value of $l$. The degeneracy $d_n$ will therefore be

$$d_n = 1 + 3 + 5 + 7 + \cdots + (2n - 1) = n^2 \qquad (6\text{-}122)$$

When $l = 0$ we say that we have a $s$ state of the system; when $l = 1$ we have a $p$ state; when $l = 2$ we have a $d$ state; when $l = 3$ we have an $f$ state; etc. This notation has historical significance only — the letters were used by spectroscopists to designate certain series of spectral lines; $s$ for *sharp*, $p$ for *principal*, $d$ for *diffuse* and $f$ for *fundamental*.

Let us now examine the nature of the wavefunctions for the hydrogen atom. In order to do this we shall discuss the radial and angular parts separately.

First we consider the radial part $R(r)$, given by Eq. (6-115). In Fig. 6-6 we have plotted this function against distance from the nucleus for the first few values of $n$ and $l$. We should note that the number of nodes for a given function is $n - l - 1$ (not including the zero values at the origin and at infinity); also we should note that $s$ functions are non-zero at the origin.

For many purposes it is useful to know the probability of finding an electron at a certain distance $r$ from the nucleus. We recall that it is the square of a wavefunction that is related to the probability. If we fix the

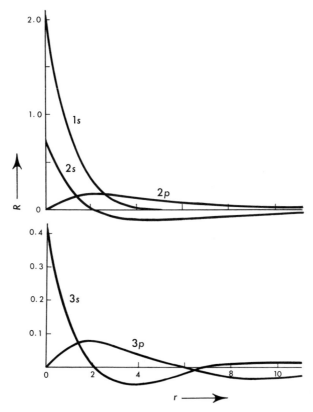

**Fig. 6-6**  The radial functions for the first few states of the hydrogen atom. ($r$ is in units of $a_0$, the radius of the first orbit in the Bohr theory of the hydrogen atom.)

values of $\theta$ and $\phi$, this determines a line emanating from the origin (i.e., from the nucleus). Having done this, we consider any particular wavefunction and examine the probability function $R^2 dr$—this gives the relative probability of finding the electron in the length $dr$ located at different places on the line. To determine the probability of the electron being in a spherical shell of thickness $dr$ at a distance $r$ from the nucleus all we have to do is integrate over the angular coordinates:

$$P(r)dr = \int_0^\pi \int_0^{2\pi} R^2 r^2 \sin\theta \, dr d\theta d\phi$$

$$= 4\pi r^2 R^2 dr \tag{6-123}$$

$P(r)$ is called the *radial distribution function*. In Fig. 6-7 we have plotted $r^2 R^2$ against $r$ for the first few values of $n$ and $l$. The significance of these curves, therefore, is that they give the relative probabilities for finding the electron at various distances from the nucleus. If we look at the curve for the 1s function, we see that the maximum probability occurs at $a_0$, the radius of the first orbit in the old Bohr theory of the hydrogen atom.

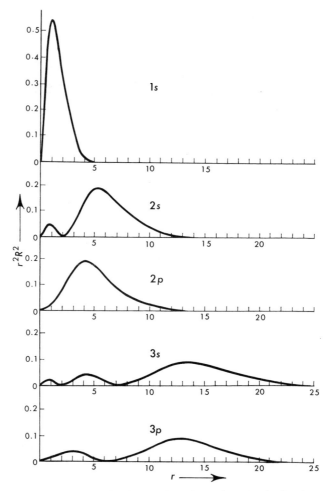

**Fig. 6-7**    Plot of $r^2 R^2$ against $r$, the distance from the nucleus, for the first few states of the hydrogen atom. ($r$ is in units of $a_0$.)

However, we should remember that the Bohr orbits were plane circular whereas we are here dealing with spherical shells of probability.

We now consider the angular parts of the wavefunctions. By so doing we can obtain the familiar pictures of orbitals. The angular parts of the functions are given by Eq. (6-102), i.e., the spherical harmonics $Y_{l,m}$ $(\theta, \phi)$. If we look at the detailed mathematical forms of these as given in the total wavefunctions Eq. (6-119), we see that for $m \neq 0$ they contain imaginary parts. It is sometimes more convenient to have real functions instead of these complex ones. To illustrate how we can replace the complex functions by real functions we consider the functions $\Psi_{2,1,1}$ and $\Psi_{2,1,-1}$. These are degenerate, since they have the same value for $n$, so we can obtain other functions corresponding to the same energy by taking linear combinations (cf. Section 4-2). Considering only the angular parts of the functions, we can examine the following linear combinations:

$$\frac{1}{2}(Y_{1,1} + Y_{1,-1}) = \left(\frac{3}{8\pi}\right)^{1/2} \sin \theta \left(\frac{e^{i\phi} + e^{-i\phi}}{2}\right)$$

$$= \left(\frac{3}{8\pi}\right)^{1/2} \sin \theta \cos \phi \qquad (6\text{-}124)$$

$$\frac{1}{2i}(Y_{1,1} - Y_{1,-1}) = \left(\frac{3}{8\pi}\right)^{1/2} \sin \theta \left(\frac{e^{i\phi} - e^{-i\phi}}{2i}\right)$$

$$= \left(\frac{3}{8\pi}\right)^{1/2} \sin \theta \sin \phi \qquad (6\text{-}125)$$

Using these relationships, and remembering that for hydrogen $Z = 1$, we can replace the three $2p$ functions in Eq. (6-119) by

$$\Psi_{2p_x} = \frac{1}{4\sqrt{2\pi}} \left(\frac{1}{a_0}\right)^{3/2} \frac{r}{a_0} \exp\left(-\frac{r}{2a_0}\right) \sin \theta \cos \phi$$

$$\Psi_{2p_y} = \frac{1}{4\sqrt{2\pi}} \left(\frac{1}{a_0}\right)^{3/2} \frac{r}{a_0} \exp\left(-\frac{r}{2a_0}\right) \sin \theta \sin \phi \qquad (6\text{-}126)$$

$$\Psi_{2p_z} = \frac{1}{4\sqrt{2\pi}} \left(\frac{1}{a_0}\right)^{3/2} \frac{r}{a_0} \exp\left(-\frac{r}{2a_0}\right) \cos \theta$$

Note that we have labeled the functions $2p_x$, $2p_y$, and $2p_z$ since these are the orbitals with these designations. We can use similar arguments to replace with real functions the other functions in Eq. (6-119) containing imaginary parts — this would give us functions corresponding to $3p_x$, $3p_y$, $3d_{xz}$, $3d_{xy}$, $3d_{yz}$, and $3d_{x^2-y^2}$ orbitals.

Knowledge about the angular parts of the wavefunctions allows us to draw the familiar contour surface diagrams of orbitals. One way of depicting orbitals is to show contour surfaces that enclose a major proportion of the electron probability density. We shall not pursue this matter further since you will be very familiar with diagrams of this type, which can be found in any elementary chemistry text. However, we should just mention that it is the square of the angular part of the wave-function that is related to the probability – this is everywhere positive. The angular function itself, for some orbitals, can be either positive or negative for different regions of space. For $s$ orbitals the angular function is everywhere positive. To illustrate how negative values can occur we can consider the $2p_z$ orbital, (cf. Fig. 6-8). Obviously, the function is

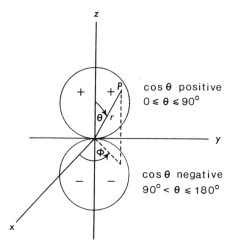

**Fig. 6-8**    Diagram illustrating how the $2p_z$ function has different signs in different regions of space. The point $P$ is located on the contour surface for the function.

positive above the $xy$ plane since here $\cos \theta$ is positive (cf. the form of the function in Eq. (6-126)); below the plane $\cos \theta$ is negative, so the function is negative. The signs of wavefunctions are important when discussing the symmetry of orbitals, and also in considerations of chemical bonding when we have to discuss the overlap of orbitals.

## SUMMARY

1. In this chapter we examined three very simple systems for which the Schroedinger equations can be solved exactly – the rigid rotator, the harmonic oscillator and the hydrogen atom.

2. In all three cases we were able to solve the equations by using a method for solving differential equations in which the solution is expressed in terms of a power series.

3. The rigid rotator Schroedinger equation gave us *spherical harmonics* $Y_{l,m}(\theta, \phi)$ as eigenfunctions. We wrote the solutions $\Psi_{l,m}$ as

$$\Psi_{l,m}(\theta, \phi) = Y_{l,m}(\theta, \phi)$$

$$= \left[\frac{(2l+1)}{4\pi} \frac{(l-|m|)!}{(l+|m|)!}\right]^{1/2} P_l^{|m|} (\cos \theta) \, e^{im\phi} \qquad (6\text{-}62)$$

$l$ and $m$ are integers and $m = -l, (-l+1), \ldots, -1, 0, 1, \ldots, l$; $P_l^{|m|}$ $(\cos \theta)$ are *associated Legendre polynomials*.

4. The solution to the harmonic oscillator problem involved the *Hermite polynomials* $H_v(y)$. We wrote the normalized solution functions $\Psi_v(y)$ as

$$\Psi_v(y) = \left(\frac{1}{\pi}\right)^{1/4} (2^v v!)^{-1/2} H_v(y) \exp\left(-\frac{y^2}{2}\right) \qquad (6\text{-}88)$$

where $y = \sqrt{\beta} x$ and $\beta = (km)^{1/2}/\hbar$ ($k$ is the *force constant* for the oscillator; $m$ is the mass of the oscillating particle).

The energies corresponding to the $\Psi_v(y)$ are given by

$$E_v = (v + \tfrac{1}{2})\hbar\omega \qquad (6\text{-}92)$$

$\omega$ is the angular frequency.

5. The hydrogen-like atom equation gave us *associated Laguerre polynomials* $L_{n+l}^{2l+1}$ $(2Zr/na_0)$ and spherical harmonics $Y_{l,m}(\theta, \phi)$ as eigenfunctions. We wrote the complete solutions $\Psi_{n,l,m}$ as

$$\Psi_{n,l,m} = -\left[\left(\frac{2Z}{na_0}\right)^3 \frac{(n-l-1)!}{2n[(n+l)!]^3}\right]^{1/2} \left(\frac{2Zr}{na_0}\right)^l \exp\left(-\frac{Zr}{na_0}\right) L_{n+l}^{2l+1}\left(\frac{2Zr}{na_0}\right)$$

$$\times Y_{l,m}(\theta, \phi) \qquad (6\text{-}118)$$

$n$ and $l$ are integers $(n > l)$; $m = 0, \pm 1, \pm 2, \ldots, \pm l$.

The energies corresponding to the $\Psi_{n,l,m}$ are given by

$$E_n = -\frac{me^4 Z^2}{2n^2 \hbar^2} \qquad (6\text{-}120)$$

## EXERCISES

**6-1**   Write down the first four Legendre polynomials, i.e., $P_0(x)$ to $P_3(x)$. Use these to form the corresponding associated Legendre polynomials.

**6-2**   Calculate the values for the first four rotational levels (i.e., for $l = 1, 2, 3,$ and

4) for the molecules $H_2$ and HF. (Hint: Consider each of the molecules to be a rigid rotator consisting of the two nuclei held together by a rigid bond. Ignore the electrons in this calculation.)

**6-3**   Verify, by substituting into Eq. (6-82), that the Hermite polynomials in Eq. (6-89) are solutions of Hermite's equation.

**6-4**   The *Laguerre polynomials of degree r* are given by

$$L_r(x) = e^x \frac{d^r}{dx^r} (x^r e^{-x})$$

The associated Laguerre polynomials of degree $r - s$ are given by

$$L_r^s(x) = \frac{d^s}{dx^s} L_r(x)$$

Determine the first few associated Laguerre polynomials that occur in Eq. (6-118).

**6-5**   Use the results of the last exercise to verify the forms of the first few functions in Eq. (6-119).

**6-6**   Show that the $1s$ and $2s$ functions for the hydrogen atom are orthogonal; also, show that the $1s$ and $2p_z$ functions are orthogonal.

**6-7**   Consider a hydrogen atom in its ground state. Determine: (a) the probability that the electron will be between $0.5a_0$ and $a_0$; (b) beyond $3a_0$.

**6-8**   Determine the mean value for the potential energy of the $1s$ electron in the hydrogen atom. (Hint: Use Eq. (4-41).)

# 7

# Approximation Methods
# in Quantum Chemistry

## 7-1 INTRODUCTION

In the last chapter we solved the Schroedinger equations for some very simple systems of chemical interest. We were able to solve the equations exactly only because the systems we chose were so simple. For more complex systems we can still set up Schroedinger equations, but we cannot usually solve these equations exactly. Therefore, we are forced to resort to approximation techniques. However, the situation is not so hopeless as it may at first seem. Though the approximation methods we use often lead to very considerable computation, modern computers allow us to handle these computations reasonably painlessly. Consequently, we are able to say that for the smaller chemical systems approximation methods give results that are quite satisfactory. What we mean by this is that the approximation methods are sufficiently refined for us to be justified in claiming that they give solutions that closely approach the exact solutions.

There are two general approximation methods for solving quantum mechanical problems. In this chapter we look at the theory behind these methods so that we shall understand what we are doing when we use them in later chapters. Other approximation methods have been used, but they are not so general in application — we shall not consider these here. The most important of the two methods we shall consider makes use of a theorem called the variation theorem — it is therefore known as the *variation method*. The other method is called *perturbation theory*. Which of these two methods we use for a particular problem naturally depends on the nature of the problem. We shall indicate the conditions under which each is most useful. Later in the book we shall illustrate their use with specific examples.

## 7-2  VARIATION METHOD

The variation method allows us to determine approximate wavefunctions and energies for our chemical systems. It is based on a very important theorem called the *variation theorem*. The form in which we shall use this theorem is as follows: If we choose an arbitrary wavefunction to describe a system, the energy calculated from this function will be an upper limit to the energy calculated from the true ground state wavefunction. (The theorem can be applied to excited states in order to find approximate wavefunctions and energies for these, but the application in this case is much more difficult.) On the basis of this we can say that if we have several approximate functions to choose from, we should choose the one that gives the lowest energy. Again, if we can find a wavefunction for a system in terms of one or more variable parameters, the values of the parameters that correspond to the lowest energy will give the best function of that type. Bearing all this in mind, what we have to do is find a general method for applying the theorem to problems of interest in chemistry. In this section we first state the theorem in mathematical form; then we prove it; finally, we briefly indicate how it is used.

But before we do all this, we should at least mention one important point to keep in mind. The variation theorem can lead us to the best function of a chosen kind where we have used the energy as a criterion for determining this function; unfortunately, however, this does not necessarily imply that the function is a good one with regard to the use of other physical properties as criteria. Nevertheless, this does not seriously detract from the importance of the variation theorem in quantum chemistry, since in chemistry, energy is the observable we are usually concerned with. In fact, the variation method has become so important in chemical theory that the variation theorem in the form we shall now examine it has become almost as important as the Schroedinger equation itself as a basic equation of quantum chemistry.

The mathematical form of the variation theorem is

$$\frac{\int \Psi^* \hat{H} \Psi \, d\tau}{\int \Psi^* \Psi \, d\tau} > E_0 \tag{7-1}$$

$E_0$ is the energy of the lowest energy state for the correct Hamiltonian $\hat{H}$; $\Psi$ is the approximation to the true wavefunction. (The only condition on $\Psi$ is that it must be normalizable over the space of the true wave-function for the system.) By expressing the theorem in this form, we are making allowance for the possibility that $\Psi$ may not be normalized. If, in fact, $\Psi$ is normalized, the denominator of the left side of Eq. (7-1) is one, and the theorem reduces to

$$\int \Psi^* \hat{H} \Psi d\tau > E_0 \qquad (7\text{-}2)$$

We now prove Eq. (7-1): The eigenfunctions of $\hat{H}$ form a complete set (cf. Section 4-7). We shall label the functions of this set $\phi_i$. Therefore, $\Psi$ can be expressed as a linear combination of the $\phi_i$ (cf. Section 4-8):

$$\Psi = c_1\phi_1 + c_2\phi_2 + c_3\phi_3 + \cdots + c_i\phi_i + \cdots$$

$$= \sum_i c_i\phi_i \qquad (7\text{-}3)$$

Substitute Eq. (7-3) into Eq. (7-1); we get

$$\frac{\int \left(\sum_i c_i\phi_i\right)\hat{H}\left(\sum_j c_j\phi_j\right)d\tau}{\int \left(\sum_i c_i\phi_i\right)\left(\sum_j c_j\phi_j\right)d\tau} > E_0 \qquad (7\text{-}4)$$

Here we have assumed that the functions are entirely real. We now recall that the complete set of eigenfunctions of $\hat{H}$ are orthonormal (cf. Section 4-7). This means that $\int \phi_i^2 d\tau = 1$; $\int \phi_i\phi_j d\tau = 0$. The denominator of the left side of Eq. (7-4) therefore becomes $\sum_i c_i^2$. Also, we remember that $\hat{H}\phi_i = E_i\phi_i$; this reduces the numerator of the left side of Eq. (7-4) to $\sum_i c_i^2 E_i$. Therefore, we need to show that

$$\frac{\sum_i c_i^2 E_i}{\sum_i c_i^2} > E_0$$

i.e.,

$$\sum_i c_i^2 E_i > E_0 \sum_i c_i^2$$

or

$$\sum_i c_i^2 (E_i - E_0) > 0 \qquad (7\text{-}5)$$

This is obviously true since $E_0$ is the lowest eigenvalue of $\hat{H}$ by definition. Therefore, Eq. (7-1) must be true.

To use the variation theorem in chemical problems, we need to guess a starting function. However, we cannot in general expect the chosen function to be very close to the true function for the system. We would therefore like to have some systematic method whereby we can modify the original function in order to obtain the best function of that type, i.e., the one leading to the lowest calculated energy. A method for doing this is to guess a function with one or more variable parameters in it. We can then adjust the values of the parameters to obtain the best function of the general form chosen. However, this is a good point to note that the final function we get by doing this can only be as good as the accuracy of our original guess at its general form.

To illustrate the application of the variation theorem as described in the last paragraph, we shall very briefly outline how it can be applied to the very important problem of chemical bonding. A major method for dealing theoretically with chemical bonding is the *Linear Combination of Atomic Orbitals or LCAO* method. The basis for this method is to assume that the wavefunctions describing electrons in molecules can be approximated by combining the atomic orbital functions of the separated atoms that form the molecules. This results in *molecular orbital functions*. To obtain these latter functions we can apply the variation theorem to trial functions that are linear combinations of the atomic orbital functions. The general form of a molecular orbital function $\Psi$ obtained in this way is therefore

$$\Psi = a_1\phi_1 + a_2\phi_2 + a_3\phi_3 + \cdots = \sum_i a_i\phi_i \qquad (7\text{-}6)$$

$\phi_i$ are the atomic orbital functions. The specific atomic orbitals to be included in $\Psi$ are determined by the particular problem and the degree of accuracy that we wish to aim for, e.g., if we were dealing with the hydrogen molecule ion $H_2^+$ — a problem we examine in detail in Chapter 10 — we could use just the two $1s$ orbitals of the separated hydrogen atoms. The $a_i$ are the linear coefficients. We now use $\Psi$ as a trial function for the application of the variation theorem. First, we determine an expression for the energy. This expression will, of course, contain the coefficients $a_i$ which can be varied to minimize the energy. The values of the $a_i$ that lead to the lowest energy determine the best function of the type Eq. (7-6).

We have not here gone into the details of the application of the variation method. However, what we have said should be sufficient to indicate how the variation theorem can be applied systematically. In later chapters we examine the use of the theorem in chemistry in more detail. In fact, we shall find it cropping up very frequently when we examine some of the applications of quantum theory to chemical problems.

## 7-3  PERTURBATION THEORY

Perturbation theory is a useful way of finding approximate functions for systems when we know the exact wavefunctions for similar systems. Two examples will illustrate what we mean by this: 1. We know the eigenfunctions for a hydrogen atom when it is not affected by external fields; provided the effect is not too great, perturbation theory allows us to approximate the new wavefunctions when the atom is put into an electrostatic field. 2. We can determine the eigenfunctions for an oscillator that is strictly harmonic. A system of interest to us that approximates the harmonic oscillator is the diatomic molecule. This system is not exactly harmonic, but perturbation theory lets us use the results of the harmonic oscillator problem to determine approximate eigenfunctions for it. What we do is add an appropriate term to the Hamiltonian for the harmonic oscillator.

We now consider the basic theory that gives us results in problems of the type outlined in the last paragraph. We shall deal only with perturbations that are independent of time. This is not to imply that time-dependent perturbations are unimportant. On the contrary, the very important field of spectroscopy makes frequent reference to them when considering the effect of external fields on a chemical system; nevertheless, we shall not deal with them here. Also, in order not to introduce complications at this stage, we shall consider only systems with non-degenerate energy levels.

We consider a system for which we know the wavefunctions $\Psi_i{}^0$ and corresponding energies $E_i{}^0$. These functions satisfy the equation

$$\hat{H}^0\Psi_i{}^0 = E_i{}^0\Psi_i{}^0 \tag{7-7}$$

Now suppose there is a small perturbation that changes the functions to $\Psi_i$; the energies change to $E_i$. Let the Hamiltonian for the perturbed system be $\hat{H}$. We can then write

$$\hat{H}\Psi_i = E_i\Psi_i \tag{7-8}$$

We now consider that any of the $\Psi_i$ can be expanded in terms of the set $\Psi_i{}^0$; this we can do because the $\Psi_i{}^0$ form a complete set (cf. Section 4-7). At the same time we recognize that as the perturbation tends to zero $\Psi_i$ tends to $\Psi_i{}^0$. We can therefore write

$$\Psi_i = \Psi_i{}^0 + a_{i1}\Psi_1{}^0 + a_{i2}\Psi_2{}^0 + \cdots + a_{ii-1}\Psi_{i-1}^0 + a_{ii+1}\Psi_{i+1}^0 + \cdots$$
$$= \Psi_i{}^0 + \sum_{j \neq i} a_{ij}\Psi_j{}^0 \tag{7-9}$$

Substitute Eq. (7-9) into Eq. (7-8); we get

$$\hat{H}\Psi_i^0 + \sum_{j \neq i} a_{ij}\hat{H}\Psi_j^0 = E_i\Psi_i^0 + E_i \sum_{j \neq i} a_{ij}\Psi_j^0 \tag{7-10}$$

Rearrange this to get

$$(\hat{H} - E_i)\Psi_i^0 + \sum_{j \neq i} a_{ij}(\hat{H} - E_i)\Psi_j^0 = 0 \tag{7-11}$$

We now write the Hamiltonian as the sum of two parts—the unperturbed Hamiltonian $\hat{H}^0$ and a perturbation term $\hat{H}'$; i.e.,

$$\hat{H} = \hat{H}^0 + \hat{H}' \tag{7-12}$$

Substitute Eq. (7-12) into Eq. (7-11); we get

$$(\hat{H}^0 + \hat{H}' - E_i)\Psi_i^0 + \sum_{j \neq i} a_{ij}(\hat{H}^0 + \hat{H}' - E_i)\Psi_j^0 = 0 \tag{7-13}$$

Now remember that (from Eq. (7-7))

$$\hat{H}^0\Psi_i^0 = E_i^0\Psi_i^0 \qquad \hat{H}^0\Psi_j^0 = E_j^0\Psi_j^0$$

This allows us to write Eq. (7-13) as

$$(E_i^0 + \hat{H}' - E_i)\Psi_i^0 + \sum_{j \neq i} a_{ij}(E_j^0 + \hat{H}' - E_i)\Psi_j^0 = 0 \tag{7-14}$$

Multiply Eq. (7-14) from the left by $\Psi_i^{0*}$ and integrate over all space; we get

$$E_i^0 \int \Psi_i^{0*}\Psi_i^0 d\tau + \int \Psi_i^{0*}\hat{H}'\Psi_i^0 d\tau - E_i \int \Psi_i^{0*}\Psi_i^0 d\tau$$

$$+ \sum_{j \neq i} E_j^0 a_{ij} \int \Psi_i^{0*}\Psi_j^0 d\tau + \sum_{j \neq i} a_{ij} \int \Psi_i^{0*}\hat{H}'\Psi_j^0 d\tau$$

$$- \sum_{j \neq i} E_i a_{ij} \int \Psi_i^{0*}\Psi_j^0 d\tau = 0 \tag{7-15}$$

Because of the orthonormality of the $\Psi_i^0$ (these are eigenfunctions of a Hamiltonian) we can write

$$\int \Psi_i^{0*}\Psi_i^0 d\tau = 1 \qquad \int \Psi_i^{0*}\Psi_j^0 d\tau = 0$$

Therefore, Eq. (7-15) becomes

$$E_i^0 - E_i + H_{ii}' + \sum_{j \neq i} a_{ij}H_{ij}' = 0 \tag{7-16}$$

Here, for convenience, we have used a notation you will often see in quantum mechanical writings; it is

$$H_{ij}' = \int \Psi_i^{0*}\hat{H}'\Psi_j^0 d\tau \tag{7-17}$$

Now multiply Eq. (7-14) from the left by $\Psi_k^{0*}$ where $k \neq i$. We shall not detail the mathematical manipulations because the principles involved are those we have just used. However, we get

$$a_{ik}(E_k{}^0 - E_i) + H'_{ki} + a_{ik}H'_{kk} + \sum_{j \neq i,k} a_{ij}H'_{kj} = 0 \qquad (7\text{-}18)$$

We now consider that the energies for the perturbed system, $E_i$, and the coefficients $a_{ij}$ can be written in terms of series in which successive terms become smaller; i.e.,

$$E_i = E_i{}^0 + E'_i + E''_i + \cdots \qquad (7\text{-}19)$$

$$a_{ij} = a'_{ij} + a''_{ij} + \cdots \qquad (7\text{-}20)$$

Here a single prime denotes a first-order term; a double prime denotes a second-order term; etc. If we substitute Eqs. (7-19) and (7-20) into Eqs. (7-16) and (7-18), we can thereby get very useful expressions. First substitute into Eq. (7-16):

$$E_i{}^0 - (E_i{}^0 + E'_i + E''_i + \cdots) + H'_{ii} + \sum_{j \neq i} a'_{ij}H'_{ij} + \sum_{j \neq i} a''_{ij}H'_{ij} + \cdots = 0 \quad (7\text{-}21)$$

We can now take out the first-order terms (remember that two first-order terms multiplied together constitute a second-order term, and so on); we get

$$E'_i = H'_{ii} = \int \Psi_i{}^{0*}\hat{H}'\Psi_i{}^0 d\tau \qquad (7\text{-}22)$$

This is the first important result—it is the first-order perturbation to the energy. We notice that it has the form of the average energy of a system for which the wavefunction is $\Psi_i{}^0$; the operator, however, is just the part of the Hamiltonian corresponding to the perturbation.

To obtain the first-order correction to the coefficients we substitute Eqs. (7-19) and (7-20) into Eq. (7-18); we get

$$H'_{ki} + a'_{ik}(H'_{kk} + E_k{}^0 - E_i{}^0 - E'_i - E''_i - \cdots) + (\text{terms that are at least}$$
$$\text{second order}) = 0 \qquad (7\text{-}23)$$

The first-order terms here give

$$H'_{ki} + a'_{ik}(E_k{}^0 - E_i{}^0) = 0 \qquad (7\text{-}24)$$

If we rearrange this, we get the expression for the first-order correction to the coefficients:

$$a'_{ik} = \frac{H'_{ki}}{E_i{}^0 - E_k{}^0} \qquad (7\text{-}25)$$

The last expression we shall obtain is the one for the second-order perturbation to the energy. This we do by picking out the second-order terms from Eq. (7-21); we get

$$-E_i'' + \sum_{j\neq i} a_{ij}' H_{ij}' = 0 \tag{7-26}$$

If we substitute for $a_{ij}'$ from Eq. (7-25) (replace $k$ by $j$), we get

$$E_i'' = \sum_{j\neq i} \frac{H_{ji}' H_{ij}'}{E_i^0 - E_j^0} \tag{7-27}$$

We could, of course, continue to examine Eqs. (7-21) and (7-23); this would give us expressions for higher-order perturbations. However, it is not usual to consider perturbations higher in order than second; certainly we shall not do so in this book.

There are a couple of points we should make before we leave the subject of perturbation theory:

1. If we substitute the expression we got for the coefficients, Eq. (7-25), into Eq. (7-9), we can write an expression for the wavefunction corrected to the first order; it is

$$\Psi_i = \Psi_i^0 + \sum_{k\neq i} \frac{H_{ki}'}{E_i^0 - E_k^0} \Psi_k^0 \tag{7-28}$$

From this we see that for $\Psi_i$ to be a good approximation, the second term in Eq. (7-28) must be small compared to $\Psi_i^0$. This is true if $H_{ki}'$ is much less than the spacing between the unperturbed energy levels. The criterion for justification when using perturbation theory is therefore

$$|H_{ki}'| \ll |E_i^0 - E_k^0| \tag{7-29}$$

2. If degeneracy occurs so that $E_i^0 = E_k^0$, the theory obviously breaks down. We shall not here consider perturbation theory for degenerate states.

## SUMMARY

1. Our purpose in this chapter was to present the two major approximation methods that we have to resort to in order to solve Schroedinger equations for all but the very simplest systems.

2. The *variation theorem* says that any approximate wavefunction for a system leads to a higher energy than the true function; in mathematical terms:

$$\frac{\int \Psi^* \hat{H} \Psi \, d\tau}{\int \Psi^* \Psi \, d\tau} > E_0 \tag{7-1}$$

$\Psi$ is an approximation to the true wavefunction; $E^0$ is the energy of the lowest state associated with the Hamiltonian $\hat{H}$.

3. *Perturbation theory* is used when we know the exact wavefunction for a similar system to the one for which we cannot find exact solutions. The important equations of perturbation theory are:

The first-order perturbation to the energy is given by

$$E_i' = H_{ii}' = \int \Psi_i^{0*} \hat{H}' \Psi_i^0 d\tau \qquad (7\text{-}22)$$

$\hat{H}'$ is the perturbation term that we have to add to the Hamiltonian for the unperturted system; $\Psi_i^0$ are the wavefunctions for the unperturbed system.

The second-order perturbation to the energy is given by

$$E_i'' = \sum_{j \neq i} \frac{H_{ji}' H_{ij}'}{E_i^0 - E_j^0} \qquad (7\text{-}27)$$

$E_i^0$ and $E_j^0$ are the energies of the unperturbed states of the system.

## EXERCISES

**7-1**  What forms would Eqs. (7-1) and (7-2) take if the trial function $\Psi$ happened to be the exact wavefunction?

**7-2**  Use the trial function

$$\phi = \exp(\lambda r^2)$$

where $\lambda$ is a variation parameter, to determine the best ground state energy for the hydrogen atom obtainable from a function of this type.

**7-3**  Consider a particle in a one-dimensional box of length $a$ (cf. Chapter 3). Assume that the wavefunction for the state of lowest energy has the form $(2/a)^{1/2} \sin \alpha x$, and use the variation theorem to determine the best value for $\alpha$.

**7-4**  Develop the theory of Section 7-3 and determine an expression for the second-order perturbation to the wavefunctions.

# 8

# Symmetry in Chemistry

## 8-1 INTRODUCTION

This chapter is, in a way, a digression from the main development of the book. Nevertheless, in it we introduce some ideas of considerable importance when we use quantum mechanics to deal with chemical problems. Very briefly, what we are going to do here is see what use we can make of the symmetry that exists in many systems of chemical interest.

If you stop and think for a moment, you will realize that many molecules and other systems of chemical interest have high degrees of symmetry. Now there is a mathematical discipline that is concerned with symmetry; it is called *group theory*. Can we use group theory in chemical problems? Yes, we can. We do not *have* to do so because all the problems that lend themselves to treatment with group theory can in theory be solved otherwise. But group theory often simplifies the problems considerably. When you recall the mathematics involved when dealing with such very simple systems as the hydrogen atom, you will appreciate that anything we can do to simplify our problems should be of major concern to us.

To give some idea of the applications of group theory in chemistry, here is a list of some of them:

1. Wavefunctions can be labeled according to their symmetry properties.

2. Molecular-orbital theory, one of the major theories for treating chemical bonding, can make considerable use of symmetry. Because molecular orbitals spread over the whole molecule, it is possible to relate their symmetry to the symmetry of the molecule—this can lead to considerable simplification in calculations.

3. It should be fairly obvious that degeneracy of energy levels is intimately related to the symmetry properties of atoms and molecules: group theory often allows us to predict degeneracy; it also allows us to examine the effect of external fields of various symmetries on degeneracy.

4. Another application that should be fairly obvious is in dealing with the normal modes of vibration of molecules.

In this chapter we shall first briefly consider the main ideas of group theory. We shall then consider, in a general way and with examples, how this theory is applied to the symmetry that exists in molecules. Finally, we shall briefly indicate the relationship between group theory and quantum chemical problems.

## 8-2  THE ELEMENTS OF GROUP THEORY

We start by defining a *group*. In everyday usage a group is a collection of things. In mathematics a group is also a collection of things or *elements*, but it is a collection with special properties. Before we list these properties we shall give an example of a mathematical group, which we can then use to illustrate the properties. The example we choose is the set of positive and negative integers including zero. In this group there are an infinite number of elements; we say that this is an *infinite group*. Normally in chemistry we are concerned with *finite groups*, i.e., those with finite numbers of elements. We call the number of elements in a group the *order* of the group.

Here are the defining properties for a group:

1. There must be a rule for combining elements of the group. When this rule is applied to any two elements the result will also be an element of the group; i.e.,

$$AB = C$$

$A$ and $B$ are elements of the group; $C$ is also an element. Though we write $AB$ as though the elements were to be multiplied together, we do not necessarily mean this. $AB$ means that the elements are to be combined according to the rule for the group.

For our example, the rule for combination is addition; e.g., $-3 + 5 = 2$, which is another element of the group.

2. If $A$ is an element of the group, there is an element $E$, which we call the *identity element*, such that

$$AE = EA = A$$

The identity element in our example is zero; e.g., $7 + 0 = 0 + 7 = 7$.

3. The associative law for combination of the elements must hold:

$$A(BC) = (AB)C$$

i.e., we must be able to combine the elements in any sequence and get the same result.

To illustrate this with our example: $3 + (2 + 4) = (3 + 2) + 4$.

4. Every element must have an *inverse* which is also an element of the group. For element $A$ we symbolize the inverse $A^{-1}$; it is defined by

$$AA^{-1} = A^{-1}A = E$$

In our example we get, for example: $5 + (-5) = (-5) + 5 = 0$; i.e., $-5$ is the inverse of $+5$.

Before we leave the definition of a group we should mention that the commutative law does not necessarily hold; i.e., $AB$ is not in general equal to $BA$. Therefore, we must be careful to keep the order of the elements when we combine them. If in fact the commutative law does hold, we say that the group is *Abelian*. Our example is an Abelian group because, for example: $5 + 4 = 4 + 5$.

## 8-3 MOLECULAR SYMMETRY OPERATIONS

Many molecules have a certain amount of symmetry that can be associated with *symmetry elements*. In order to indicate how we can classify these symmetry elements we consider, as an example, the water molecule. We shall examine its symmetry properties and show that these lend themselves to the use of group theory. The molecule is shown in Fig. 8-1.

If we rotate the molecule by 180° about the axis labeled $C_2$, the new configuration is equivalent to the original one; we say that we have performed a symmetry operation on the molecule. Another symmetry operation for the water molecule is reflection in the plane labeled $\sigma_v$; this operation also gives an equivalent configuration to the original one. Similarly, we can reflect the molecule through the $\sigma'_v$ plane. Since the molecule is totally in this plane, this operation must be a symmetry operation also. There is yet another symmetry operation for the water molecule — leave the molecule as it is. This last operation may seem quite trivial, but it is an important one from the point of view of application of group theory to the molecule; it corresponds to the identity element, which is necessary for a set of elements to constitute a group.

These four symmetry operations for the water molecule correspond to

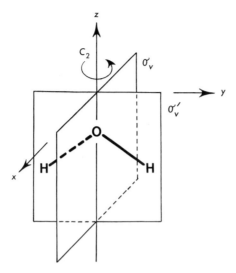

**Fig. 8-1**   The symmetry operations for the water molecule (point group $C_{2v}$).

symmetry elements that form a group. The law for combination of the elements is to perform the symmetry operations successively in the order given. Let us, by way of some examples, show that these symmetry elements do in fact form a group.

The combination requirement for the elements to form a group can be illustrated as follows: Reflect the molecule through the $\sigma_v$ plane; then rotate it through 180° about the $C_2$ axis. This is equivalent to leaving it as it is, i.e., the operation corresponding to the identity element. You can easily check for yourself that the combination requirement holds for all the elements.

We have already mentioned that the identity element for the group corresponds to leaving the molecule alone. A requirement of the identity element is that it must commute with all the other elements and leave them unchanged. Obviously, the operation of leaving the molecule alone satisfies this requirement.

We can check the associative law by considering the following: Reflect the molecule through the $\sigma_v$ plane; then rotate it about the $C_2$ axis—this brings it back to its original position. Now reflect it through the $\sigma_v'$ plane — this is equivalent to leaving it alone. Now instead of performing the operations in this order, consider what happens when the combined operation of rotation and reflection through the $\sigma_v'$ plane is performed after the operation of reflection through the $\sigma_v$ plane: the result is the same.

This, therefore, illustrates that the associative law holds in this case. You can easily check that it holds in all other cases.

The inverse of any of the symmetry elements, for this particular molecule, is the same element. (We should note that this is not generally true for groups associated with molecular symmetry elements.) As an example, consider the following: Rotate the molecule through 180°. Now rotate it again in the same direction by 180° — this brings it back to its original position. The total result of the two operations is therefore the same as if we had left the molecule alone, i.e., the identity operation. It is very obvious in this case that the order of performing the operations does not matter. In this case then, the second rotation operation gives the inverse element for the first rotation operation. A similar statement can be made for all the other symmetry elements.

We call the group to which the symmetry elements of the water molecule belongs a *point group*. This is because all the symmetry operations always leave some point of the molecule fixed. Point groups for molecules are given symbols (sometimes known as *Schoenflies symbols*), and each point group has associated with it a particular set of symmetry operations. The water molecule, our example, belongs to the point group $C_{2v}$. Why we give it this symbol we shall see presently. In the meantime we must realize that there are other possible molecular symmetry operations than the ones associated with the water molecule. These can be best classified as shown in Table 8-1.

**Table 8-1**   Symmetry elements of molecules.

| Symmetry element | Symbol | Symmetry operation |
|---|---|---|
| Identity | $E$ | Leave molecule alone |
| Plane of symmetry | $\sigma$ | Reflect through plane |
| Center of symmetry | $i$ | Invert through center of symmetry |
| Axis of symmetry | $C_n$ | Rotate about axis by $360/n$ degrees |
| Rotation-reflection axis of symmetry | $S_n$ | Rotate about axis by $360/n$ degrees; then reflect through plane perpendicular to axis |

Looking at this table, we can see why we labeled the axis of rotation and the planes of symmetry as we did in Fig. 8-1. The axis of rotation was a $C_2$ axis, meaning that the symmetry element corresponded to rotation by 360/2 or 180°. We labeled the planes $\sigma_v$ and $\sigma'_v$; the subscript indicates that they are vertical planes through the molecule. By vertical we mean that they contain the *principal axis*; the principal axis, by convention, is always placed vertically. To determine which axis is the principal one when there are several axes, consider the value of $n$; the principle axis is the one for which $n$ is greatest, (cf. Fig. 8-2). Sometimes we need to

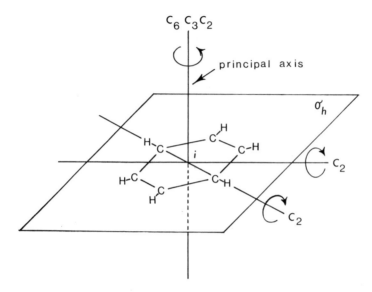

**Fig. 8-2**   Diagram illustrating the principal axis and some of the symmetry operations for the benzene molecule (point group $D_{6h}$). For clarity, not all the symmetry operations are shown.

use symmetry planes perpendicular to the principal axis; these are horizontal planes so we give them the symbol $\sigma_h$. A further type of possible symmetry plane is one that bisects the angles between $n$ two-fold axes perpendicular to the principle axis; this type of plane is labeled $\sigma_d$ (*d* for dihedral). In Fig. 8-2 we illustrate some of the symmetry operations not exemplified by Fig. 8-1.

We can now return to the consideration of the point groups to which various molecules belong. Obviously, different molecules have different combinations of symmetry elements associated with them. However, if we examined many molecules, we would realize that only certain combinations occur. We have already mentioned that these combinations are called point groups. Table 8-2 lists some representative point groups that occur in molecular problems; we have used the symbols in Table 8-1

**Table 8-2**   Some molecular point groups.

| Point group | Symmetry elements (in addition to identity element) |
|---|---|
| $C_2$ | $C_2$ |
| $C_i$ | $i$ |
| $C$ | $\sigma$ |
| $C_{2v}$ | $C_2, 2\sigma_v$ |
| $C_{3v}$ | $C_3, 3\sigma_v$ |
| $C_{\infty v}$ | $C_\infty, \infty\sigma_v$ |
| $C_{2h}$ | $i, C_2, \sigma_h$ |
| $D_{2d}$ | $3C_2$ (mutually perpendicular), $S_4$ (coincident with one $C_2$), $2\sigma_d$ (through $S_4$ axis) |
| $D_{2h}$ | $i$, $3C_2$ (mutually perpendicular), $3\sigma$ (mutually perpendicular) |
| $D_{\infty h}$ | $i, C_\infty, \infty C_2$ (perpendicular to $C_\infty$ axis) $\infty\sigma_v, \sigma_h$ |
| $T_d$ | $3C_2$ (mutually perpendicular), $4C_3, 6\sigma, 3S_4$ (coincident with $C_2$ axes) |
| $O_h$ | $i, 3C_4$ (mutually perpendicular), $4C_3, 3S_4$ and $3C_2$ (coincident with $C_4$ axes), $6C_2, 9\sigma, 4S_6$ (coincident with $C_3$) |

to tabulate the symmetry elements associated with them. When you see a figure before a symmetry element this refers to the number of times that element occurs; e.g., $3C_2$ means that there are three two-fold axes of rotation.

If we examine the entry in Table 8-2 for the $C_{2v}$ group, we shall realize why we said that the water molecule belongs to it. Actually, it would be a good idea if you now examined some common molecules and determined their point groups (Exercise 8-2 suggests some examples for you to try). When you do this it is usually best to first decide whether the molecule belongs to one of the special groups $T$, $T_d$, $O$, or $O_h$. If it does not, look for a principal axis; decide whether it is a $C_n$ or $S_n$ axis. Then look for any other symmetry elements.

We should mention here that the point groups listed in Table 8-2 are not all the possible point groups that can occur for molecules. These are, however, more than sufficient for our present needs.

## 8-4  REPRESENTATION OF GROUPS

In this section we shall see how we can conveniently deal with the symmetry elements that molecules possess. But first we should ask the question: What happens to the wavefunction for a molecule when a symmetry operation is performed on the molecule? The physical interpretation of the wavefunction is in terms of electron probability densities. Since the symmetry operation leaves the molecule in an equivalent configuration, we can say that these probability densities for the various parts of the system will remain unaltered. Now it is $\Psi^2$ that gives the probability densities; therefore, if the symmetry operation transforms $\Psi$ into $\pm \Psi$, the condition that the probability densities remain unaltered is satisfied. In operator notation we can express this:

$$\hat{R}\Psi = +1\Psi \quad \text{or} \quad \hat{R}\Psi = -1\Psi \tag{8-1}$$

$\hat{R}$ is the operator associated with a symmetry operation $R$ for the molecule; $\Psi$ is an eigenfunction of this operator. As an example we can consider the water molecule (point group $C_{2v}$); the following set of equations would be valid:

$$\hat{E}\Psi = +1\Psi \qquad \hat{\sigma}_v \Psi = +1\Psi$$
$$\hat{C}_2\Psi = +1\Psi \qquad \hat{\sigma}_v' \Psi = +1\Psi \tag{8-2}$$

We can say, therefore, that the four numbers $+1, +1, +1, +1$ characterize the symmetry of the eigenfunction.

These are not the only sets of numbers that characterize the symmetries of the eigenfunctions of the molecule. There is, however, a restriction on the sets: the numbers have to combine together in such a way that the numbers that represent the symmetry operations combine to give the number that represents the combined operation. We can easily clarify this with an example: In the water molecule the successive operations $\sigma_v'$ and $C_2$ give the same result as the operation $\sigma_v$; i.e.,

$$\hat{\sigma}_v' \hat{C}_2 \Psi = \hat{\sigma}_v \Psi \tag{8-3}$$

Equations (8-2) obviously satisfy this condition. But so will the set of numbers obtained from the equations

$$\hat{E}\Psi = +1\Psi \qquad \hat{\sigma}_v \Psi = -1\Psi$$
$$\hat{C}_2\Psi = +1\Psi \qquad \hat{\sigma}_v' \Psi = -1\Psi \tag{8-4}$$

Any set of numbers that can be assigned to the elements of a group and satisfy the rules for combining the elements is called a *representation* of the group. We list some of the representations of the $C_{2v}$ group in Table 8-3:

**Table 8-3**   Some representations of the $C_{2v}$ group.

| $E$ | $C_2$ | $\sigma_v$ | $\sigma_v'$ |
|-----|-------|-----------|-------------|
| 1 | 1 | 1 | 1 |
| 1 | 1 | $-1$ | $-1$ |
| 1 | $-1$ | 1 | $-1$ |
| 1 | $-1$ | $-1$ | 1 |

These are the only sets of numbers that fulfill the specified conditions. However, there are other ways by which we can represent the group. But before we can look at these we must digress for a while to say something about matrices and determinants. The mathematics that follows is probably not new to you. If this is the case, you can treat the section in small type as a review of what you will need to understand the remainder of this chapter; if it is not, an understanding of it will be sufficient for your present needs.

## Matrices and Determinants

A matrix is simply an array of numbers. The following are examples of matrices:

$$(a) \quad \begin{bmatrix} 8 & 5 \\ 6 & -3 \end{bmatrix} \qquad (b) \quad \begin{bmatrix} 5 & 3 \\ 2 & -2 \\ 9 & -1 \end{bmatrix}$$

The use of matrices in problems is that the numbers, or *elements*, of a matrix can be, for example, the coefficients of a set of linear equations. For example, the matrix (a) can represent the coefficients in the equations

$$8x + 5y = 0$$
$$6x - 3y = 0$$

Again, a matrix may be the coordinates of points in a coordinate system. For example, the matrix (b) can represent two points with coordinates $(5, 2, 9)$ and $(3, -2, -1)$. And there are other uses for matrices. Here we are interested in matrices with regard to using them as representations of groups.

If a set of matrices is to be a representation of a point group for a molecule, we need a law for combining them; furthermore, this law has to be in accord with the properties of the symmetry elements of the molecule. There are, in fact, mathematical rules for combining matrices. We can, for example, add them together or subtract them. The rules are merely definitions, but the definitions are such that they make matrix algebra useful in applications. The rule we are most interested in here is the one for multiplying matrices; this we shall now describe; then we shall use it in our development of the application of group theory to chemical problems.

Let us first consider an example; then we can generalize it. Consider two matrices $A$ and $B$. (We usually give matrices letter symbols to save writing them in full.)

$$A = \begin{bmatrix} a & b \\ c & d \end{bmatrix} \qquad B = \begin{bmatrix} e & f \\ g & h \end{bmatrix}$$

The product of $A$ and $B$ will be another matrix, which we shall call $C$. We write the multiplication operation

$$AB = C$$

where

$$C = \begin{bmatrix} a & b \\ c & d \end{bmatrix}\begin{bmatrix} e & f \\ g & h \end{bmatrix} = \begin{bmatrix} ae+bg & af+bh \\ ce+dg & cf+dh \end{bmatrix}$$

What we have done here is as follows: To get the element in the first row and first column of $C$ we multiplied the elements of the first row of $A$ by the corresponding elements of the first column of $B$ and added the results. To get the element in the first row and second column of $C$ we multiplied the elements of the first row of $A$ by the elements of the second column of $B$ and added the results. And so on. The general rule is this: To get the element in row $r$ and column $s$ of the product matrix, multiply the elements of the $r$th row of the first matrix by the corresponding elements of the $s$th column of the second matrix; then add the results. In symbols we write this

$$C_{rs} = \sum_t A_{rt}B_{ts} \qquad (8\text{-}5)$$

$C_{rs}$ is the element of $C$ in the $r$th row and the $s$th column.

The example we considered for matrix multiplication involved two square matrices of the same size and shape. Matrices do not have to be square; nor do the ones to be multiplied have to be of the same size and shape. However, if we consider the rule for multiplication for a moment, we shall realize that matrices cannot be multiplied unless the number of elements in the rows of the first one is equal to the number of elements in the columns of the second one.

You will have noticed that we have been careful to specify the order in which matrices are to be multiplied. This is very important since matrix multiplication is not necessarily commutative, i.e., $AB$ is not necessarily equal to $BA$. You should try some examples to convince yourself that this is so.

There is considerable special terminology associated with matrix algebra. We shall not here attempt to make even a reasonably complete summary of this. Instead we shall just summarize what we need for the rest of this chapter. Before we can do this, however, we must say something about determinants.

Any *square* matrix may have associated with it a *determinant*. The determinant of a matrix $M$ is distinguished from the matrix by writing it $|M|$. The major difference between a matrix and a determinant is that the determinant has a value—it is a scalar quantity. Matrices, remember, are merely arrays of elements.

How do we find the value of a determinant? Consider the general determinant of order $n$ (the *order* of a determinant is the number of rows or columns in it).

$$|M| \equiv \begin{bmatrix} M_{11} & M_{12} & M_{13} & \cdots & M_{1n} \\ M_{21} & M_{22} & M_{23} & \cdots & M_{2n} \\ \cdots\cdots\cdots\cdots\cdots\cdots\cdots\cdots \\ M_{n1} & M_{n2} & M_{n3} & \cdots & M_{nn} \end{bmatrix}$$

We define the *minor* of the element $M_{rs}$ as the determinant left when we cross out the row $r$ and the column $s$ containing $M_{rs}$. The signed minor we get by multiplying the minor by $(-1)^{r+s}$ we call the *cofactor* of the element. To find the value of a determinant we multiply each element of some chosen row or column (any row or column) by its cofactor and sum the results. We can best illustrate this by an example:

$$\begin{vmatrix} 2 & 5 & 0 \\ 1 & 5 & 2 \\ 0 & 1 & 0 \end{vmatrix} = 2\begin{vmatrix} 5 & 2 \\ 1 & 0 \end{vmatrix} - 5\begin{vmatrix} 1 & 2 \\ 0 & 0 \end{vmatrix} + 0\begin{vmatrix} 1 & 5 \\ 0 & 1 \end{vmatrix}$$
$$= (2 \times -2) - (5 \times 0) + (0 \times 1) = -4$$

We have considered the first row and used the fact that the value of a $2 \times 2$ determinant is given by

$$\begin{vmatrix} a & b \\ c & d \end{vmatrix} = ad - cb$$

This is known as the *Laplace development* of a determinant. For larger determinants this method reduces the order of the determinant or determinants by one each time we use it. In this case we would continue to apply the Laplace development until all the determinants were second order.

We can now return to consideration of the special terminology needed for this chapter. What we actually need to know is the names of some special matrices and how to obtain them—this is summarized in Table 8-4. It would be well for you to practice forming these special matrices from a given matrix.

**Table 8-4**    Special matrices.

| Name of matrix | Notation | Method for obtaining from $M$ |
|---|---|---|
| $M$ transpose or transpose of $M$ | $\tilde{M}$ | Interchange rows and columns |
| Complex conjugate of $M$ (usually just called the conjugate) | $M^*$ | Replace each element of $M$ by its complex conjugate |
| Transpose conjugate or associate matrix (improperly called the adjoint matrix in quantum mechanics — do not confuse this with the next matrix) | $M\dagger$ | Replace each element of $M$ by its complex conjugate, then transpose; i.e., form $\tilde{M}^*$ |
| Adjoint or adjugate of $M$ | $\hat{M}$ | Replace each element of $M$ by its cofactor, then transpose |
| Inverse of $M$ | $M^{-1}$ | Divide each element of $\hat{M}$ by $\|M\|$; i.e., $M^{-1} = \hat{M}/\|M\|$ |

This is all we need to know about matrices and determinants for this chapter; anything else we need in the remaining chapters we shall introduce at the time when we want it.

Now let us see what happens if we try to use matrices to form a representation of a group. We said earlier that any set of numbers that can be assigned to the elements of a group and satisfy the rules for combining the elements is a representation of the group. In fact, the definition of a representation is more general than this since sets of matrices can be assigned to the elements of the group to form representations. We can continue to use the water molecule as an example of this generalization. Consider, for instance, the following set of matrices corresponding to the symmetry elements $E, C_2, \sigma_v$, and $\sigma'_v$ of the $C_{2v}$ group:

$$
\begin{matrix}
E & C_2 & \sigma_v & \sigma'_v \\
\begin{bmatrix} 1 & 0 \\ 0 & 1 \end{bmatrix} & \begin{bmatrix} -1 & 0 \\ 0 & -1 \end{bmatrix} & \begin{bmatrix} 1 & 0 \\ 0 & -1 \end{bmatrix} & \begin{bmatrix} -1 & 0 \\ 0 & 1 \end{bmatrix}
\end{matrix}
\qquad (8\text{-}6)
$$

Is this a representation of the group? If it is, we should be able to multiply any two of the matrices and get the matrix corresponding to the symmetry element we would have obtained by combining the symmetry elements corresponding to the two matrices that were multiplied. This condition is indeed satisfied, as we can easily check. Consider the following as an example: We saw before that

$$
\sigma'_v C_2 = \sigma_v \qquad (8\text{-}7)
$$

If the set in Eq. (8-6) is a representation of the group, then

$$
\begin{bmatrix} -1 & 0 \\ 0 & 1 \end{bmatrix} \begin{bmatrix} -1 & 0 \\ 0 & -1 \end{bmatrix} = \begin{bmatrix} 1 & 0 \\ 0 & -1 \end{bmatrix}
$$

This is true.

There are other sets of matrices that can form representations of the group. In fact, there are an infinite number. However, although there are so many representations, some of them are especially important; these we call irreducible representations. How these arise we see as follows.

Remember that the representations of a group are matrices. (If the representations are single numbers, this is just the special case where the matrices have just one element in them.) Often it is possible to break down these matrices into smaller matrices that also form representations of the group. To break down a representation in this way we perform what is known as a *similarity transformation* on it. A similarity transformation on a matrix $A$ is one such that

$$
X^{-1}AX = B \qquad (8\text{-}8)
$$

$X$ is any matrix and $X^{-1}$ is its inverse. $B$ is then said to be the *similarity transform* of $A$. Now if the matrices $A, B, C, \ldots$ form a representation of the group, we may be able to break these down by similarity transformations so that the new matrices have smaller sub-matrices on the diagonals and zeros elsewhere. What we mean is it may be possible to do the following:

$$X^{-1}AX = A' = \begin{bmatrix} A_1 & 0 & 0 \\ 0 & A_2 & 0 \\ 0 & 0 & A_3 \end{bmatrix}$$

$$X^{-1}BX = B' = \begin{bmatrix} B_1 & 0 & 0 \\ 0 & B_2 & 0 \\ 0 & 0 & B_3 \end{bmatrix}$$

(8-9)

etc.

$A_1, B_1, C_1, \ldots$ are matrices of the same dimension. We can easily prove that $A_1, B_1, C_1, \ldots$ form a new representation of the group; also that $A_2, B_2, C_2 \ldots$ and $A_3, B_3, C_3, \ldots$ form representations of the group. We shall now do this.

Let us suppose that one of the requirements of the group is

$$AB = C \tag{8-10}$$

We shall first prove that

$$A'B' = C' \tag{8-11}$$

Substitute from Eq. (8-9) into Eq. (8-11); we get

$$X^{-1}AXX^{-1}BX = X^{-1}CX \tag{8-12}$$

Now,

$$XX^{-1} = I$$

where $I$ is a unit matrix, i.e., a square matrix with ones as diagonal elements and zeros elsewhere:

$$I = \begin{bmatrix} 1 & 0 & 0 & \cdots \\ 0 & 1 & 0 & \cdots \\ 0 & 0 & 1 & \cdots \\ \cdots & \cdots & \cdots & \cdots \end{bmatrix}$$

Therefore, Eq. (8-12) gives us

$$X^{-1}ABX = X^{-1}CX \tag{8-13}$$

This is because multiplication of a matrix by a unit matrix of the same dimension leaves it unchanged (you can easily check this). Now multiply both sides of Eq. (8-13) from the left by $X$ and from the right by $X^{-1}$.

Since $XX^{-1} = X^{-1}X = I$, we get

$$AB = C$$

We have proved, therefore, that writing $A'B' = C'$ is equivalent to writing $AB = C$. Therefore, the new matrices $A'$, $B'$, $C'$, ... form a representation of the group. But since matrices are equal only if their corresponding elements are equal, the rules for matrix multiplication require that the smaller matrices $A_1$, $B_1$, $C_1$, ..., $A_2$, $B_2$, $C_2$, ..., etc. must combine in the following way:

$$\begin{aligned} A_1B_1 &= C_1 \\ A_2B_2 &= C_2 \end{aligned} \tag{8-14}$$

etc.

We can clarify this last statement best by giving an example. Consider the following matrix product:

$$\begin{bmatrix} 1 & 1 & 0 & 0 & 0 & 0 \\ 0 & 2 & 0 & 0 & 0 & 0 \\ 0 & 0 & 3 & 1 & 1 & 0 \\ 0 & 0 & 2 & 0 & 4 & 0 \\ 0 & 0 & 1 & 1 & 3 & 0 \\ 0 & 0 & 0 & 0 & 0 & 2 \end{bmatrix} \begin{bmatrix} 2 & 3 & 0 & 0 & 0 & 0 \\ 0 & 1 & 0 & 0 & 0 & 0 \\ 0 & 0 & 2 & 2 & 1 & 0 \\ 0 & 0 & 1 & 1 & 0 & 0 \\ 0 & 0 & 0 & 3 & 0 & 0 \\ 0 & 0 & 0 & 0 & 0 & 3 \end{bmatrix} = \begin{bmatrix} 2 & 4 & 0 & 0 & 0 & 0 \\ 0 & 2 & 0 & 0 & 0 & 0 \\ 0 & 0 & 7 & 10 & 3 & 0 \\ 0 & 0 & 4 & 16 & 2 & 0 \\ 0 & 0 & 3 & 12 & 1 & 0 \\ 0 & 0 & 0 & 0 & 0 & 6 \end{bmatrix}$$

The product matrix is obviously blocked out in the same way as the original matrices. Furthermore, the elements in any particular block in the product matrix are determined solely by the corresponding blocks in the original matrices. A little thought will convince you that this must be generally true: when we multiply two matrices that are blocked out in the same way, the corresponding blocks in each may be considered independently. In our example this means that we obtained the product matrix by the following multiplications:

$$\begin{bmatrix} 1 & 1 \\ 0 & 2 \end{bmatrix} \begin{bmatrix} 2 & 3 \\ 0 & 1 \end{bmatrix} = \begin{bmatrix} 2 & 4 \\ 0 & 2 \end{bmatrix}$$

$$\begin{bmatrix} 3 & 1 & 1 \\ 2 & 0 & 4 \\ 1 & 1 & 3 \end{bmatrix} \begin{bmatrix} 2 & 2 & 1 \\ 1 & 1 & 0 \\ 0 & 3 & 0 \end{bmatrix} = \begin{bmatrix} 7 & 10 & 3 \\ 4 & 16 & 2 \\ 3 & 12 & 1 \end{bmatrix}$$

$$[2][3] = [6]$$

Referring back to Eq. (8-11), we see that the matrix product $A'B'$ is of the type we have just been discussing. Therefore, Eq. (8-14) is true, which means that $A_1, B_1, C_1, \ldots, A_2, B_2, C_2, \ldots$, and so on, are representations of the group.

When a representation of a group can be broken down by similarity transformations, we say that we have a *reducible representation*. On the other hand, if we can find no matrix $X$ that will reduce all the matrices of the representation, we say that we have an *irreducible representation*. Much of the application of group theory to chemical problems involves finding the irreducible representations of groups.

## 8-5 CHARACTERS OF REPRESENTATIONS AND CHARACTER TABLES

Let us consider some representations of the $C_{3v}$ group, to which molecules such as ammonia and methyl chloride belong. These are set out in Table 8-5.

Representations (1), (2), and (3) are irreducible representations; (4) is a reducible representation. If we inspect (4), we quickly realize that it is made up of the representations (1), (2), and (3) — we say that (4) is reducible to (1), (2), and (3). Here it is obvious how we should break down the reducible representation to irreducible representations. If the reducible representation were not in diagonal form (i.e., having the irreducible representations nicely set out on the diagonal with zeros elsewhere) it would be much less easy to break it down; we would have to perform the necessary similarity transformations — this for large matrices is a tedious job, though computers take much of the work out of it. However, fortunately, it is not usually necessary to do this since the information we need can usually be obtained from the characters of the representations.

The *character* of a representation is the sum of its diagonal elements (in mathematics this is usually called the *trace* of a matrix), i.e., for the matrix

$$\begin{bmatrix} A_{11} & A_{12} & A_{13} & \cdots \\ A_{21} & A_{22} & A_{23} & \cdots \\ A_{31} & A_{32} & A_{33} & \cdots \\ \cdots\cdots\cdots\cdots\cdots \end{bmatrix}$$

the character is $A_{11} + A_{22} + A_{33} + \cdots$. We give the character the symbol $\chi$. We give the symbol $\chi(R)$ to the character of a representation of the $R$th symmetry operation. If we wish to distinguish a particular representation, we use a subscript — the character of the $R$th symmetry operation in the $i$th representation is therefore written $\chi_i(R)$.

**Table 8-5** Representations for the $C_{3v}$ group.

| | $E$ | $C_3$ | $C_3^2$ | $\sigma_{va}$ | $\sigma_{vb}$ | $\sigma_{vc}$ |
|---|---|---|---|---|---|---|
| (1) | $1$ | $1$ | $1$ | $1$ | $1$ | $1$ |
| (2) | $1$ | $1$ | $1$ | $-1$ | $-1$ | $-1$ |
| (3) | $\begin{bmatrix} 1 & 0 \\ 0 & 1 \end{bmatrix}$ | $\begin{bmatrix} -\frac{1}{2} & \frac{\sqrt{3}}{2} \\ -\frac{\sqrt{3}}{2} & -\frac{1}{2} \end{bmatrix}$ | $\begin{bmatrix} -\frac{1}{2} & -\frac{\sqrt{3}}{2} \\ \frac{\sqrt{3}}{2} & -\frac{1}{2} \end{bmatrix}$ | $\begin{bmatrix} -1 & 0 \\ 0 & 1 \end{bmatrix}$ | $\begin{bmatrix} \frac{1}{2} & -\frac{\sqrt{3}}{2} \\ -\frac{\sqrt{3}}{2} & -\frac{1}{2} \end{bmatrix}$ | $\begin{bmatrix} \frac{1}{2} & \frac{\sqrt{3}}{2} \\ \frac{\sqrt{3}}{2} & -\frac{1}{2} \end{bmatrix}$ |
| (4) | $\begin{bmatrix} 1 & 0 & 0 \\ 0 & 1 & 0 \\ 0 & 0 & 1 \end{bmatrix}$ | $\begin{bmatrix} 1 & 0 & 0 \\ 0 & -\frac{1}{2} & \frac{\sqrt{3}}{2} \\ 0 & -\frac{\sqrt{3}}{2} & -\frac{1}{2} \end{bmatrix}$ | $\begin{bmatrix} 1 & 0 & 0 \\ 0 & -\frac{1}{2} & -\frac{\sqrt{3}}{2} \\ 0 & \frac{\sqrt{3}}{2} & -\frac{1}{2} \end{bmatrix}$ | $\begin{bmatrix} 1 & 0 & 0 \\ 0 & -1 & 0 \\ 0 & 0 & 1 \end{bmatrix}$ | $\begin{bmatrix} 1 & 0 & 0 \\ 0 & \frac{1}{2} & -\frac{\sqrt{3}}{2} \\ 0 & -\frac{\sqrt{3}}{2} & -\frac{1}{2} \end{bmatrix}$ | $\begin{bmatrix} 1 & 0 & 0 \\ 0 & \frac{1}{2} & \frac{\sqrt{3}}{2} \\ 0 & \frac{\sqrt{3}}{2} & -\frac{1}{2} \end{bmatrix}$ |

An important property of characters of representations is the following: A character of a representation is unchanged by a similarity transformation. In our example (Table 8-5) it is obvious that

$$\chi_4(R) = \chi_1(R) + \chi_2(R) + \chi_3(R) \tag{8-15}$$

But is this true when the matrices are not in diagonal form? Yes, it is, and we can easily prove it:

Consider two matrices $A$ and $B$ related by the similarity transformation

$$A = X^{-1}BX \tag{8-16}$$

We wish to show that

$$\chi_A = \chi_B \tag{8-17}$$

To do this we first need to prove that if $C = AB$ and $D = BA$ where $A, B, C,$ and $D$ are matrices, then the characters of $C$ and $D$ are equal: A diagonal element in $C$ is given by $C_{jj}$. Therefore, the character of $C$ is given by

$$\chi_C = \sum_j C_{jj} \tag{8-18}$$

Referring to Eq. (8-5), we can write any element $C_{rs}$ in the matrix product $AB$ as $\sum_t A_{rt}B_{ts}$. Therefore,

$$\chi_C = \sum_j \sum_k A_{jk}B_{kj} \tag{8-19}$$

For the same reasons,

$$\chi_D = \sum_k D_{kk} = \sum_k \sum_j B_{kj}A_{jk} \tag{8-20}$$

But,

$$\sum_k \sum_j B_{kj}A_{jk} = \sum_j \sum_k A_{jk}B_{kj} = \chi_C \tag{8-21}$$

We can now return to the original problem. From Eq. (8-16) we get

$$\chi_A = \chi \quad \text{of} \quad X^{-1}BX \tag{8-22}$$

Now the associative law holds for matrix multiplication, a fact you can easily check using some examples. Therefore, using this fact and the result we just obtained, we get

$$\begin{aligned} \chi \quad \text{of} \quad X^{-1}BX &= \chi \quad \text{of} \quad (X^{-1}B)X \\ &= \chi \quad \text{of} \quad X(X^{-1}B) = \chi \quad \text{of} \quad (XX^{-1})B \\ &= \chi_B \end{aligned} \tag{8-23}$$

We have, therefore, proved what we wanted — that the character of a matrix remains the same after a similarity transformation. So far as we are concerned, what this means is that for a particular symmetry operation $R$ the sum of the characters of all the irreducible representations we obtain from a reducible representation is equal to the character of the reducible representation. In mathematical language we write this

$$\chi(R) = \sum_i a_i \chi_i(R) \tag{8-24}$$

$a_i$ is the number of times the $i$th irreducible representation occurs in the reducible representation.

Since it is the characters of representations that are most useful to us, it is usual to list these in what we call *character tables*. We show the character table for the $C_{3v}$ group in Table 8-6. You will see that all we

**Table 8-6**    Character table for the $C_{3v}$ group.

| $C_{3v}$ | $E$ | $2C_3$ | $3\sigma_v$ |
|---|---|---|---|
| $A_1$ | 1 | 1 | 1 |
| $A_2$ | 1 | 1 | −1 |
| $E$ | 2 | −1 | 0 |

have done is added up the diagonal elements of the representation matrices. (Remember that the first two representations are just one-dimensional matrices.) For other character tables you can refer to such books as reference F2 in the Bibliography.

Some explanation of Table 8-6 is needed. If we look at Table 8-5, we see that the characters for the $C_3$ and $C_3{}^2$ operations are the same for a particular representation; also, the characters for $\sigma_{va}$, $\sigma_{vb}$, and $\sigma_{vc}$ are the same. We say that the operations $C_3$ and $C_3{}^2$ belong to the same *class*; the operations $\sigma_{va}$, $\sigma_{vb}$, and $\sigma_{vc}$ belong to another class. For convenience we write in the character table $2C_3$ and $3\sigma_v$. The operation $E$ belongs to a class of its own and is therefore shown separately. We can, in fact, make the general statement that characters associated with operations of the same class are the same for any particular representation.

The definition of a class is: A class is a set of elements of a group such that

$$X^{-1}AX = B$$

$X$ is any element of the group; $A$ is any member of the class, and $B$ is also a member of the same class; i.e., this operation, when performed on a member of the class, must give another member of the class.

As an example of the last statement we can consider the class in the $C_{3v}$

group associated with the operations $\sigma_{va}$, $\sigma_{vb}$, and $\sigma_{vc}$:

$$E^{-1}\sigma_{va}E = E\sigma_{va}E = \sigma_{va}$$

$$C_3^{-1}\sigma_{va}C_3 = C_3^2\sigma_{va}C_3 = \sigma_{vc}$$

$$\sigma_{va}^{-1}\sigma_{vb}\sigma_{va} = \sigma_{va}\sigma_{vb}\sigma_{va} = \sigma_{vc}$$

and so on.

One further point about classes: The number of elements in a class is always an integral factor of the order of the group. We can see that this is true for the present case—the order of the group is six, and the number of elements in the classes are: one for $E$, two for $C_3$ and three for $\sigma_v$.

Look at the character table again. We see some symbols in the first column. These are generally used symbols and mean as follows:

1. $A$ and $B$ are used to label irreducible representations of dimension one. Here we have two such representations. $A$ is used when the character of the highest rotation (i.e., having the largest value of $n$ in $C_n$) is $+1$ (i.e., symmetric with respect to this rotation); $B$ when this is $-1$ (i.e., antisymmetric with respect to this operation). Here the rotations in the one-dimensional representation both have characters $+1$ so we have two to label $A$; we distinguish them with subscripts. There are no one-dimensional representations with characters for the highest rotation $-1$, so we have no $B$'s in the table.

$E$ is used to label irreducible representations of dimension two. There is one such representation associated with the $C_{3v}$ group.

Other symbols are used to label higher-dimensioned representations: $T$ is used for three-dimensional representations; $G$ is used for four-dimensional representations; etc.

2. The meanings of subscripts are as follows:

1 is used if a $\sigma_v$ operation is symmetric.
2 is used if a $\sigma_v$ operation is antisymmetric.
$g$ is used if a $i$ operation is symmetric.
$u$ is used if a $i$ operation is antisymmetric.

3. The meanings of primes on the symbols in character tables are as follows: A single prime is used when a $\sigma_h$ operation is symmetric; a double prime is used when a $\sigma_h$ operation is antisymmetric.

We therefore see in character tables such symbols as $A_g$, $B_{1u}$, $T_{2g}$, and $A''$.

## 8-6  PROPERTIES OF IRREDUCIBLE REPRESENTATIONS AND THEIR CHARACTERS

The properties we are going to list here are very useful in applications of group theory to chemistry. We shall not. however. prove these properties, but merely state them with some examples.

1. *The sum of the squares of the dimensions of the irreducible representations of a group is equal to the order of the group.*

As an example of this we can consider the $C_{3v}$ group, the irreducible representations for which we gave in Table 8-5. Remember that we labeled these irreducible representations $A_1, A_2$, and $E$; they have dimensions one, one, and two. Therefore,

$$[d(A_1)]^2 + [d(A_2)]^2 + [d(E)]^2 = 6$$

Six is indeed the order of the $C_{3v}$ group.

2. *The characters of the irreducible representations of a group behave like orthogonal vectors in a h-dimensional space where h is the order of the group.*

If we label the character of the $R$th symmetry operation in the $i$th irreducible representation $\chi_i(R)$, this means that

$$\sum_R \chi_i(R)\chi_j(R) = 0 \qquad \text{if } i \ne j \qquad (8\text{-}25)$$

(You should note that if we were dealing with complex characters. this equation would read

$$\sum_R [\chi_i(R)]^*\chi_j(R) = 0 \qquad (8\text{-}26)$$

Similar adjustments would have to be made to the equations that follow.)

Once again we can use the $C_{3v}$ group for illustration. Consider the second and third representations. i.e.. $A_2$ and $E$:

$$\sum_R \chi_i(R)\chi_j(R) = \sum_R \chi_{A_2}(R)\chi_E(R)$$

$$= (1 \times 2) + 2(1 \times -1) + 3(-1 \times 0) = 0$$

If $i = j$. then

$$\sum_R [\chi_i(R)]^2 = h \qquad (8\text{-}27)$$

As an example of this, we consider the $E$ representation of the $C_{3v}$ group; we get

$$\sum_R [\chi_E(R)]^2 = 2^2 + 2(-1)^2 + 3(0)^2 = 6$$

It is useful to combine Eqs. (8-25) and (8-27); we then get

$$\sum_R \chi_i(R)\chi_j(R) = h\delta_{ij} \qquad (8\text{-}28)$$

where $\delta_{ij}$ is the Kronecker delta.

Equation (8-28) is a most important equation. It expresses a necessary and sufficient condition that a representation is irreducible.

3. We now derive a very important equation. Equation (8-24) was

$$\chi(R) = \sum_i a_i\chi_i(R)$$

Multiply this by $\chi_j(R)$ and sum over all operations; we get

$$\sum_R \chi(R)\chi_j(R) = \sum_i \sum_R a_i\chi_i(R)\chi_j(R) \qquad (8\text{-}29)$$

Now substitute Eq. (8-28) into the right side of Eq. (8-29); we get

$$\sum_R \chi(R)\chi_j(R) = ha_j \qquad (8\text{-}30)$$

(Note that $a_j$ is the only remaining coefficient because the right side of Eq. (8-29) is zero if $i \neq j$); or

$$a_j = \frac{1}{h} \sum_R \chi(R)\chi_j(R) \qquad (8\text{-}31)$$

We can illustrate this result very easily. Suppose we have found, some-how or other, a reducible representation for the $C_{3v}$ group with characters given by

| $C_{3v}$: | E | $2C_3$ | $3\sigma_v$ |
|---|---|---|---|
| Γ: | 3 | 0 | 1 |

Referring to the character table for the $C_{3v}$ group (Table 8-6), and using Eq. (8-31), we get

$$a_{A_1} = \tfrac{1}{6}[\,(1 \times 3 \times 1) + (2 \times 0 \times 1) + (3 \times 1 \times 1)\,] = 1$$

$$a_{A_2} = \tfrac{1}{6}[(1 \times 3 \times 1) + (2 \times 0 \times 1) + (3 \times 1 \times -1)] = 0$$

$$a_E = \tfrac{1}{6}[(1 \times 3 \times 2) + (2 \times 0 \times -1) + (3 \times 1 \times 0)] = 1$$

What this means is that the reducible representation can be broken down into irreducible representations; $A_1$ and $E$ each occur once, whilst $A_2$ does not occur in this particular representation. This, then, gives us a method for finding the number of times each irreducible representation occurs in a reducible representation — often in problems this is all we need to know.

## 8-7  RELATIONSHIP OF GROUP THEORY TO QUANTUM CHEMISTRY

As we intimated at the beginning of the chapter, the use of group theory in chemistry is that it can greatly simplify many problems. Here we indicate why this is so.

If we perform a symmetry operation on a molecule, the resulting configuration is exactly equivalent to the original configuration—this is the meaning of a symmetry operation. But what does this mean in quantum mechanical terms? It means that the electron probability densities must remain unaltered and that the energy and Hamiltonian for the system must not change. The relationship between these three is given by the basic eigenvalue equation

$$\hat{H}\Psi_i = E_i\Psi_i$$

$E_i$ is the energy corresponding to the eigenfunction $\Psi_i$.

If a symmetry operation is performed on the system, we get

$$\hat{R}\hat{H}\Psi_i = \hat{R}E_i\Psi_i \tag{8-32}$$

But since $\hat{R}$ does not affect $\hat{H}$ or $E$, we can write

$$\hat{H}(\hat{R}\Psi_i) = E_i(\hat{R}\Psi_i) \tag{8-33}$$

The function $\hat{R}\Psi_i$ is therefore an eigenfunction of $\hat{H}$ with the same eigenvalue as $\Psi_i$. We can therefore conclude that, if the state is non-degenerate,

$$\hat{R}\Psi_i = \pm\,\Psi_i \tag{8-34}$$

for normalized functions.

What if the state is degenerate with two or more eigenfunctions, which we shall label $\Psi_{in}$, corresponding to the given energy? Then the energy can remain the same under the symmetry operation provided the original eigenfunction is transformed into a linear combination of the degenerate functions (cf. Section 4-2). In general, therefore, we can write for an $l$-fold degenerate state

$$\hat{R}\Psi_{in} = \sum_{m=1}^{l} r_{mn}\Psi_{im} \tag{8-35}$$

Here the $r_{mn}$ are the coefficients of the linear combination. They form representation matrices expressing the effect of the symmetry operations on the set of degenerate eigenfunctions $\Psi_{in}$. We say that the $\Psi_{in}$ are a *basis* for the representation of the group since the representation is generated by the application of the symmetry operations on the set of degenerate functions. In fact, the representation is irreducible. We shall not prove this last statement for the general case where degeneracy occurs. However, it is evident for the special case of non-degenerate eigenfunctions. Looking at Eq. (8-34), we see that by applying the symmetry operations of a group to an eigenfunction belonging to a non-degenerate eigenvalue we have generated a representation of the group in which each matrix is one-dimensional with the single element $\pm 1$ — obviously this is an irreducible representation.

This result is very important in quantum chemical problems, because it relates the eigenfunctions of a molecule to its symmetry properties. In fact, it limits the forms of the eigenfunctions a symmetrical molecule can have. When we are looking for an unknown eigenfunction this can be very useful information.

## 8-8 THE DIRECT PRODUCT

There is one more concept we should examine before leaving our discussion on the basic theory of molecular symmetry: Sometimes we come across functions that are products of other functions and we would like to know the transformation properties of the product functions.

Consider two sets of functions $A_1, A_2, \ldots, A_r$ and $B_1, B_2, \ldots, B_s$ that are bases for representations of a point group of a molecule. We assume that we know the transformation properties of these functions. If we have a symmetry operation $R$, we can write the effect of this on the sets of functions as follows (cf. Section 8-7):

$$\hat{R}A_i = \sum_{j=1}^{r} a_{ji}A_j \qquad (8\text{-}36)$$

$$\hat{R}B_k = \sum_{l=1}^{s} b_{lk}B_l \qquad (8\text{-}37)$$

Now consider the $rs$ product functions $A_iB_k$. This set of functions is called the *direct product*; it forms a basis for a representation of the group which, in general, will be a reducible representation. We can obtain a very important theorem concerning the character of this representation — this we now do.

Consider the effect of the symmetry operation on the product functions;

it is given by

$$\hat{R}A_iB_k = \sum_{j=1}^{r} \sum_{l=1}^{s} a_{ji} b_{lk} A_j B_l$$

$$= \sum_{j=1}^{r} \sum_{l=1}^{s} c_{jl,ik} A_j B_l \tag{8-38}$$

Here we have written $c_{jl,ik}$ for the direct product matrix to distinguish it from the matrix we get by ordinary matrix multiplication. The order of this matrix is $rs$. A simple example of a direct product matrix is

$$\begin{bmatrix} a_{11} & a_{12} \\ a_{21} & a_{22} \end{bmatrix} \times \begin{bmatrix} b_{11} & b_{12} \\ b_{21} & b_{22} \end{bmatrix} = \begin{bmatrix} a_{11}b_{11} & a_{11}b_{12} & a_{12}b_{11} & a_{12}b_{12} \\ a_{11}b_{21} & a_{11}b_{22} & a_{12}b_{21} & a_{12}b_{22} \\ a_{21}b_{11} & a_{21}b_{12} & a_{22}b_{11} & a_{22}b_{12} \\ a_{21}b_{21} & a_{21}b_{22} & a_{22}b_{21} & a_{22}b_{22} \end{bmatrix}$$

i.e., to obtain the direct product matrix we multiply every element in the one matrix by every element in the other.

We can easily see that in this case the character of the product matrix is equal to the product of characters of the matrices that were multiplied together to obtain it. In fact, this is a general theorem — we prove it as follows:

$$\chi_{AB}(R) = \sum_{j} \sum_{l} c_{jl,jl}$$

$$= \sum_{j} \sum_{l} a_{jj} b_{ll} \tag{8-39}$$

i.e.,

$$\chi_{AB}(R) = \chi_A(R) \chi_B(R) \tag{8-40}$$

Equation (8-40) is a statement of the theorem we wished to prove. In words, this means that the character for a symmetry operation in a direct product representation is equal to the product of the corresponding characters for the component representations.

As an example of what we have just proved, consider the characters for the $C_{3v}$ group (Table 8-6). To obtain the characters for the direct product representation $A_2 \cdot E$, we just multiply the corresponding characters in the $A_2$ and $E$ representations; we get

| | $E$ | $2C_3$ | $3\sigma_v$ |
|---|---|---|---|
| $A_2 \cdot E$: | 2 | $-1$ | 0 |

These are the characters for the $E$ representation. We can obtain characters for the other direct product representations in a similar way.

## SUMMARY

1. In this chapter we considered the very basic theory by which it is possible to take advantage of the symmetry that many molecules possess.

2. We started by considering the mathematical theory called *group theory* and saw that the symmetry elements associated with molecules obey the requirements of mathematical groups.

3. Molecules with symmetry belong to *point groups*, so called because symmetry operations on a molecule leave at least one point fixed in space. We listed and named some of the more important point groups to which molecules belong.

4. A *representation of a group* is any set of numbers or matrices that satisfy the rules for combining the elements of the group (molecular symmetry elements in this case). A *reducible representation* is one that can be broken down by similarity transformations to give smaller-dimensioned representations; an *irreducible representation* is one that cannot.

5. When using group theory, we usually need to know only the characters of representations. The *character* of a representation matrix is the sum of its diagonal elements. Characters are listed for the various point groups in *character tables*.

6. The following two equations are very important:

$$\sum_R \chi_i(R)\, \chi_j(R) = h\delta_{ij} \tag{8-28}$$

$$a_j = \frac{1}{h} \sum_R \chi(R)\, \chi_j(R) \tag{8-31}$$

$\chi_i(R)$ is the character for the $R$th symmetry operation in the $i$th irreducible representation; $\chi(R)$ is the character for the $R$th operation in a reducible representation; $h$ is the order of the group, i.e., the number of elements in it; $a_j$ is the number of times the $j$th irreducible representation occurs in the reducible representation. Equation (8-31) gives us a method for determining the number of times each irreducible representation of a group occurs in a reducible representation we might have found.

7. The effect of a symmetry operation on a non-degenerate molecular wavefunction is to give back the same function or to change its sign. If the function is degenerate, the symmetry operation changes it to a linear combination of the degenerate functions.

8. The character for a symmetry operation in a *direct product representation* is equal to the product of the corresponding characters for the component representations.

## EXERCISES

**8-1** Draw a diagram of the $NH_3$ molecule and identify the symmetry elements.

**8-2** Determine the point groups for the following molecules: $CO_2$, $C_2H_4$, HCN, $SF_6$, $CH_4$.

**8-3** Show that the operations $C_3$ and $C_3{}^2$ of the $C_{3v}$ group form a class.

**8-4** Verify that the representations of the $C_{3v}$ group given in Table 8-5 are in accord with the effects of combining the various symmetry operations for the molecule.

**8-5** The matrices that can form representations of point groups for molecules all have the property of being *unitary*. What this means is that if $M$ is unitary,

$$M^{-1} = M\dagger \quad \text{or} \quad MM\dagger = I$$

Show that the representation matrices in Table 8-5 are in fact unitary.

**8-6** Use the character table for the $C_{2v}$ group (Table 8-7) to find the component irreducible representations of the following reducible representation:

| $E$ | $C_2$ | $\sigma_v$ | $\sigma_v'$ |
|---|---|---|---|
| 9 | −1 | 1 | 3 |

**Table 8-7**   Character table for the $C_{2v}$ group.

| $C_{2v}$ | $E$ | $C_2$ | $\sigma_v$ | $\sigma_v'$ |
|---|---|---|---|---|
| $A_1$ | 1 | 1 | 1 | 1 |
| $A_2$ | 1 | 1 | −1 | −1 |
| $B_1$ | 1 | −1 | 1 | −1 |
| $B_2$ | 1 | −1 | −1 | 1 |

**8-7** Show that the direct product of the $E$ representation for the $C_{3v}$ group with itself is a reducible representation. Further, determine the irreducible representations into which the reducible representation breaks down.

# 9

# Many-electron Atoms

## 9-1 INTRODUCTION

The systems we have so far looked at in detail have been very simple ones by quantum mechanical standards – they have contained just one or two particles. In these cases we have been able to solve exactly the Schroedinger equations. Chemical systems of practical interest almost always contain many particles. We could, of course, attempt to solve exactly the Schroedinger equations for these more complicated systems. Unfortunately, however, the Schroedinger equations for many-particle systems are not susceptible to exact solution, though for the simpler systems of this type, at least, approximation techniques give results very close to the exact solutions.

In this chapter we consider the special case of a many-particle system, the many-electron atom. By so doing we can illustrate some of the basic problems encountered when dealing with many-particle systems in general. At the same time we will, of course, be considering systems of very considerable chemical interest.

First we set up and examine the Schroedinger equation for a many-electron atom. We indicate how to circumvent the problems associated with the term in the Hamiltonian – the electron-electron repulsion term – that causes the mathematical difficulty. Next, by introducing the Pauli exclusion principle and the question of indistinguishability of electrons, we consider how to write wavefunctions for atoms with many electrons in them. We very briefly discuss the electronic configurations of the elements, using the quantum mechanical approximations we will have introduced to deal with many-electron atoms. After this we turn to a consideration of the total angular momentum of many-electron atoms; here we examine how to combine the orbital and spin angular momenta for these systems. Finally, we briefly consider some aspects of a theory that allows us to make quite sophisticated approximations to the solutions of some important many-electron problems.

## 9-2 THE SCHROEDINGER EQUATION FOR MANY-ELECTRON ATOMS

We start our examination of many-electron atoms by considering the Schroedinger equation for these systems. To set up this equation we need the Hamiltonian operator. For our present purposes it is sufficient if we consider the spin-free, *non-relativistic Hamiltonian* since this gives the major portion of the energy of a stationary state. We therefore include in the Hamiltonian only the kinetic energy terms for the electrons and the potential energy terms for the nucleus-electron and electron-electron interactions. For more accurate considerations we would have to take account of such phenomena as spin-orbit coupling between the spin and orbital angular momenta of the electrons, and coupling between nuclear spin and electron orbital angular momenta. Neglecting such factors and non-relativistic phenomena that are not important for our present purposes, we can write the non-relativistic Hamiltonian for a many-electron atom as (in cgs units)

$$\hat{H} = -\frac{\hbar^2}{2m} \sum_i \nabla_i^2 - \sum_i \frac{Ze^2}{r_i} + \sum_{i<j} \frac{e^2}{r_{ij}} \tag{9-1}$$

The first term is the Laplacian operator for the electrons; i.e., it refers to their kinetic energy. The other terms are potential energy terms due to coulombic interactions. The second takes into account the nucleus-electron interactions ($Ze$ is the nuclear charge; $r_i$ is the distance of electron $i$ from the nucleus). The third takes account of the electron-electron repulsions ($r_{ij}$ is the distance between electron $i$ and electron $j$. The notation $i < j$ on the summation sign means that the summation is to be made for all values of $j$, and that for each value of $j$ there is a further summation over all values of $i$ that are less than $j$—this ensures that each repulsion is taken into account only once.)

Before we continue to use the Hamiltonian in Eq. (9-1) we can simplify its form a little by using what we call *atomic units* (usually abbreviated a.u.). Atomic units save writing a lot of constants, so are very often used in quantum mechanical expressions. Furthermore, these units are defined in terms of fundamental constants such as $m$, the rest mass of the electron, and $e$, the electronic charge; therefore, if we express the results of calculations in atomic units, these results do not change if more accurate values are determined for the fundamental constants.

In atomic units, mass is expressed in terms of the rest mass of the electron. The atomic unit of length is $a_0$, the radius of the first Bohr orbit in the Bohr theory of the hydrogen atom: this unit is called the *bohr*:

$$1 \text{ bohr} = a_0 = \frac{\hbar^2}{me^2} = 0.52917 \times 10^{-8} \text{ cm} \qquad (9\text{-}2)$$

The atomic unit of time is the time required for an electron to travel one atomic unit of length in the first Bohr orbit (about $2.42 \times 10^{-17}$ sec).

The atomic unit of energy is defined by the *hartree* where

$$1 \text{ hartree} = \frac{me^4}{\hbar^2} = 27.210 \text{ eV} \qquad (9\text{-}3)$$

(Table 9-1 shows the relationship between the hartree and other units of energy that we commonly use in chemistry.) The hartree is twice the

**Table 9-1**    Conversion table for energy units†.

|  | hartree | eV | $cm^{-1}$ | ergs | kcal $mole^{-1}$ |
|---|---|---|---|---|---|
| 1 hartree | 1 | 27.21 | $2.195 \times 10^5$ | $4.359 \times 10^{-11}$ | $6.275 \times 10^2$ |
| 1 eV | $3.675 \times 10^{-2}$ | 1 | 8065 | $1.602 \times 10^{-12}$ | 23.06 |
| 1 $cm^{-1}$ | $4.556 \times 10^{-6}$ | $1.239 \times 10^{-4}$ | 1 | $1.986 \times 10^{-16}$ | $2.859 \times 10^{-3}$ |
| 1 erg | $2.294 \times 10^{10}$ | $6.242 \times 10^{11}$ | $5.034 \times 10^{15}$ | 1 | $1.439 \times 10^{13}$ |
| 1 kcal $mole^{-1}$ | $1.594 \times 10^{-3}$ | $4.336 \times 10^{-2}$ | $3.498 \times 10^2$ | $6.947 \times 10^{-14}$ | 1 |

†Based on the fundamental physical constants given by Taylor, B. N., Parker, W. H., and Langenberg, D. N., *Rev. Mod. Phys.*, **41**, 375 (1969).

ionization potential of the hydrogen atom if the reduced mass is replaced by the rest mass of the electron. Other conventions are sometimes used for expressing energy. For example, the unit of energy is sometimes defined as half the unit we have defined here (i.e., $me^4/2\hbar^2$). We can also give energy in terms of the *Rydberg constant for infinite mass* $R_\infty$ where

$$R_\infty = \frac{e^2}{2a_0hc} = 1.09737 \times 10^5 \text{cm}^{-1} \qquad (9\text{-}4)$$

or we can use the double Rydberg ($2R_\infty$) to be consistent with the use of the hartree as defined by Eq. (9-3). (The infinite mass for the Rydberg unit refers to the mass of the nucleus since the limit of the reduced mass $\mu$ as the mass of the nucleus $M_N$ tends to infinity is given by

$$\lim_{M_N \to \infty} \mu = \lim_{M_N \to \infty} \left( \frac{mM_N}{m + M_N} \right) = m )$$

Using atomic units, the Hamiltonian operator in Eq. (9-1) becomes (cf. Exercise 9-1)

$$\hat{H} = -\frac{1}{2} \sum_i \nabla_i^2 - \sum_i \frac{Z}{r_i} + \sum_{i<j} \frac{1}{r_{ij}} \tag{9-5}$$

## 9-3  THE INDEPENDENT-PARTICLE MODEL

When we set up a Schroedinger equation with the Hamiltonian of Eq. (9-5) we get a partial differential equation. Unlike our simple examples, however, we cannot solve it easily. This unhappy state of affairs is due to the electron-electron repulsion term, i.e., the third term; this prevents our using the separation of variables technique. If we attempt to separate the variables, we do not get simple ordinary differential equations to solve, as we did before.

Because of these difficulties, we are forced to use approximation techniques. One approach to the problem is to omit the offending term in the Hamiltonian, i.e., the electron-electron repulsion term. Since this is an important term in the Hamiltonian, we would not expect this solution to the problem to be a very good one. However, it does give a starting point for more satisfactory solutions, so we shall examine it in detail.

The Hamiltonian we shall work with is, therefore,

$$\hat{H}^0 = -\frac{1}{2} \sum_i \nabla_i^2 - \sum_i \frac{Z}{r_i} \tag{9-6}$$

This is now just a summation over the one-electron Hamiltonians $\hat{h}_i(i)$ where $\hat{h}_i(i)$ is given by

$$\hat{h}_i(i) = -\frac{1}{2} \nabla_i^2 - \frac{Z}{r_i} \tag{9-7}$$

When we apply the separation of variables technique to the equation

$$\hat{H}^0 \Phi^0 = E^0 \Phi^0 \tag{9-8}$$

we write $\Phi^0$ as a product function:

$$\Phi^0 = \prod_{i=1}^{N} \phi_i(i) \tag{9-9}$$

Here $\phi_i(i)$ is a function only of the coordinates of electron $i$; $N$ is the number of electrons in the atom. The equations we are left with to solve all have the form

$$\hat{h}_i\phi_i = E_i\phi_i \qquad (9\text{-}10)$$

where $E_i$ is the energy of the $i$th electron. There is one equation of this type for each electron.

Now if we refer to the form of $\hat{h}_i$, as given in Eq. (9-7), we realize that all the equations in Eq. (9-10) are hydrogen-like (cf. Section 6-5). The functions $\phi_i$ are therefore just hydrogen-like orbitals. We have given these orbital functions before (cf. Eq. (6-118)). Briefly, we write them

$$\phi_i = \phi_{n,l,m} = R_{n,l}(r)Y_{l,m}(\theta,\phi) \qquad (9\text{-}11)$$

In this approximation the total electronic energy $E$ for the system is given by the sum of the energies $E_i$, i.e.,

$$E = \sum_i E_i \qquad (9\text{-}12)$$

Before considering how to improve the independent-particle model of the many-electron atom, we shall apply the simple model to the helium atom.

## 9-4  THE HELIUM ATOM

The simplest many-electron system of chemical interest is the helium atom. For the present, we consider just the ground state of the atom; later in this chapter, after we have discussed some further general aspects of many-electron systems, we shall examine excited states. Also for the present, we shall not consider the spin of the electrons — we shall see later that this would not in any case make any difference to our present energy calculations.

We gave the non-relativistic Hamiltonian for the general case of a many-electron atom in Eq. (9-5). For the helium atom, with electrons labeled 1 and 2, we get

$$\hat{H} = -\frac{1}{2}(\nabla_1{}^2 + \nabla_2{}^2) - \frac{2}{r_1} - \frac{2}{r_2} + \frac{1}{r_{12}} \qquad (9\text{-}13)$$

The first term is, as usual, the kinetic energy term; the other three terms are the potential energy terms — these take account of nucleus-electron

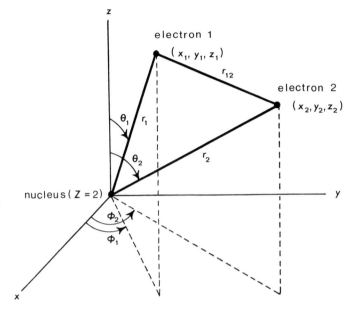

**Fig. 9-1**   Coordinate systems for the helium atom. The nucleus is at the origin.

and electron-electron interactions. The meaning of $r_1$, $r_2$, and $r_{12}$ is clarified in Fig. 9-1.

Using the theory of the last section, we omit the electron-electron repulsion term in Eq. (9-13), i.e., $1/r_{12}$. The resulting Hamiltonian can then be separated into two parts:

$$\hat{H}^0 = \hat{h}_1(1) + \hat{h}_2(2) \qquad (9\text{-}14)$$

$\hat{h}_1$ and $\hat{h}_2$ are hydrogen-like Hamiltonians that depend respectively on the coordinates of electrons 1 and 2. The Schroedinger equation with $\hat{H}^0$ as the Hamiltonian is now separable into the two equations:

$$-\frac{1}{2}\nabla_1^2\phi_1(1) - \frac{2}{r_1}\phi_1(1) = E_1\phi_1(1) \qquad (9\text{-}15a)$$

$$-\frac{1}{2}\nabla_2^2\phi_2(2) - \frac{2}{r_2}\phi_2(2) = E_2\phi_2(2) \qquad (9\text{-}15b)$$

These have the form of the hydrogen-like equation we examined in Section 6-5. We can therefore write down the solutions without further ado (cf. Eq. (6-119)):

$$\phi_1(1) = \phi_{1s}(1) = \frac{(2)^{3/2}}{\pi^{1/2}} \exp(-2r_1) \qquad (9\text{-}16a)$$

$$\phi_2(2) = \phi_{1s}(2) = \frac{(2)^{3/2}}{\pi^{1/2}} \exp(-2r_2) \qquad (9\text{-}16b)$$

We obtain the complete solution $\Phi^0$ as (cf. Eq. (9-9))

$$\Phi^0 = \frac{(2)^3}{\pi} \exp[-2(r_1 + r_2)] \qquad (9\text{-}17)$$

According to Eq. (9-12), the energy of the helium atom in this approximation is given by

$$E = E_1 + E_2$$

where $E_1 = E_2 =$ energy of the $1s$ hydrogen-like orbital for $Z = 2$. Using Eqs. (6-120) and (9-3), $E$ is therefore $-4$ hartrees. The experimental value is $-2.905$ hartrees. Obviously, this is not a very satisfactory result. We therefore ask: How might we improve the approximation?

One way to improve the calculation would be to use the approximate wavefunction Eq. (9-17) and the Hamiltonian $\hat{H}$ given in Eq. (9-13). The energy would then be given by

$$E = \int \Phi^0 \hat{H} \Phi^0 d\tau$$

$$= \int \phi_{1s}(1)\phi_{1s}(2)\left[ -\frac{1}{2}(\nabla_1{}^2 + \nabla_2{}^2) - \frac{2}{r_1} - \frac{2}{r_2} + \frac{1}{r_{12}} \right]\phi_{1s}(1)\phi_{1s}(2) d\tau \qquad (9\text{-}18)$$

The volume element $d\tau$ here refers to the coordinates of electron 1 and electron 2. Separating the integrals in Eq. (9-18), we get

$$E = -\frac{1}{2}\int \phi_{1s}(1)\phi_{1s}(2)[\nabla_1{}^2 + \nabla_2{}^2]\phi_{1s}(1)\phi_{1s}(2) d\tau$$

$$-2\int \phi_{1s}(1)\phi_{1s}(2)\left[\frac{1}{r_1} + \frac{1}{r_2}\right]\phi_{1s}(1)\phi_{1s}(2) d\tau$$

$$+\int \phi_{1s}(1)\phi_{1s}(2)\left[\frac{1}{r_{12}}\right]\phi_{1s}(1)\phi_{1s}(2) d\tau \qquad (9\text{-}19)$$

We shall not evaluate these integrals since the mathematics is rather tedious. However, we can recognize that the first two give just the energy of the independent-particle model, i.e., $-4$ hartrees. The third integral is somewhat more difficult to evaluate than the first two because of the presence of the term $r_{12}$†. Its value is 1.25 hartrees. The total energy given by this approximation is therefore $-2.75$ hartrees. This is a considerable improvement over our previous result, but it is still not particularly good. (Remember that the experimental value is $-2.905$ hartrees.) We must therefore take a further look at the problem to see if a better approximation can be made.

To determine the wavefunction Eq. (9-17) we have ignored the electron-electron repulsion term in the Hamiltonian. We realized when we did this that we were making a rather serious approximation. However, it did allow us to separate, and therefore solve, the Schroedinger equation. The question we now ask, therefore, is: Can we within this approximation somehow take account of the fact that the electrons are affected by each other's presence? We can answer this question by considering what effect the nuclear charge has on one of the electrons when there is another electron present in the atom.

If electron 1 is very far from the nucleus, we would expect that electron 2 is probably much closer; therefore, since electron 1 is almost completely screened from the nucleus by electron 2, the nuclear potential acting on it is approximately $-(2-1)/r_1$, i.e., $-1/r_1$. On the other hand, if electron 1 is very close to the nucleus, electron 2 hardly screens it at all, in which case it feels almost the full nuclear charge of 2; i.e., the nuclear potential acting on it is approximately $-2/r_1$. On the average, the *effective nuclear charge Z'*, so far as the electrons are concerned, is somewhere between 1 and 2. We might therefore expect that we would obtain better wavefunctions than those in Eq. (9-16) if instead of using the Hamiltonians in Eq. (9-15), we used Hamiltonians with $Z'$ substituted for the actual nuclear charge of 2. We could then determine the best value for $Z'$ by applying the variation theorem.

Using this method, the total wavefunction, analogous to Eq. (9-17), is

$$\Phi^{0\prime} = \frac{(Z')^3}{\pi} \exp\left[-Z'(r_1+r_2)\right] \tag{9-20}$$

We now use this wavefunction and the Hamiltonian in Eq. (9-13) to obtain an expression for the energy in terms of $Z'$; it is

$$E' = (Z')^2 - \frac{27}{8} Z' \tag{9-21}$$

All we have to do now is apply the variation theorem to determine the

†For details of the evaluation of the integrals in Eq. (9-19) you can consult reference C2 in the Bibliography, pp. 179–181.

value of $Z'$ that causes the energy to be a minimum. Differentiating Eq. (9-21) with respect to $Z'$, we get

$$\frac{dE'}{dZ'} = 2Z' - \frac{27}{8} = 0 \qquad (9\text{-}22)$$

Therefore, $Z' = 27/16$. Substituting this value of $Z'$ into Eq. (9-21), we get $E' = -2.848$ hartrees. This compares quite well, considering the approximations we have made, with the experimental value of $-2.905$ hartrees.

The actual nuclear charge for the helium atom is 2. The results we have just obtained suggest that the effective nuclear charge for the electrons is 1.69. The use of an effective nuclear charge has given a reasonably satisfactory result for the ground state energy of the atom because it recognizes that each electron is affected by the presence of the other. We could, in fact, use this method for atoms with more than two electrons, in which case we would assign an effective nuclear charge $Z'_i$ for each electron $i$. By so doing we would partially account for the error introduced by omitting the electron-electron repulsion term from the Hamiltonian. However, this is a rather naïve approach to the problem of electron-electron interactions. Later in this chapter we briefly examine a more sophisticated solution to the problem, the Hartree-Fock self-consistent field method.

## 9-5  SPIN-ORBITALS

So far in this discussion of many-electron atoms we have said nothing about the spin angular momentum of the electrons. We introduced the subject of electron spin in Section 5-4. There we discussed the spin quantum number $m_s$ and said that there are two possible spin functions that can be used to describe the intrinsic angular momentum of a single electron. We labeled these two functions by $\alpha$ (corresponding to $m_s = +\frac{1}{2}$) and $\beta$ (corresponding to $m_s = -\frac{1}{2}$).

To write the complete wavefunction for an electron we have to combine its space function with its spin function. The result is a *spin-orbital* which is a function of both the space and spin coordinates of the electron. To be more explicit, a spin-orbital $\phi(x,y,z,s)$ is a product of the space function $\phi(x,y,z)$ (assuming the coordinate system is cartesian) and the spin function ($\alpha$ or $\beta$); i.e.,

$$\psi(x,y,z,s) = \phi(x,y,z,)\alpha \quad \text{or} \quad \phi(x,y,z,)\beta \qquad (9\text{-}23)$$

To consider a specific example, we can use the case of an electron in a one-dimensional box. The spin-orbitals $\psi_{n_x, m_s}$ are then (cf. Eq. (3-23))

$$\psi_{n_x, 1/2}(x, s) = \phi_{n_x}(x)\alpha = \left(\frac{2}{a}\right)^{1/2} \sin\left(\frac{n_x \pi}{a} x\right)\alpha \qquad (9\text{-}24a)$$

$$\psi_{n_x, -1/2}(x, s) = \phi_{n_x}(x)\beta = \left(\frac{2}{a}\right)^{1/2} \sin\left(\frac{n_x \pi}{a} x\right)\beta \qquad (9\text{-}24b)$$

## 9-6  INDISTINGUISHABLE PARTICLES: THE PAULI EXCLUSION PRINCIPLE

The next point to consider is how to take account of electron spin when writing total wavefunctions for many-electron atoms. This leads us into a discussion of the symmetry properties of systems of indistinguishable particles. We shall first consider a general system of indistinguishable particles, then look in more detail at the specific case of systems of electrons.

Consider a system of identical and indistinguishable particles (e.g., electrons or protons). We can describe the system with a wavefunction $\Psi(q_i, s_i)$, a function of the spatial $q_i$ and spin $s_i$ coordinates of the system. The question we ask is: What happens to $\Psi$ if we exchange the coordinates of two of the particles? We can answer the question by considering the physical significance of $\Psi$, i.e., $|\Psi|^2$. Since the particles are indistinguishable, the operation of exchanging their coordinates cannot alter $|\Psi|^2$. However, this leaves two possibilities for the effect on $\Psi$ itself. If we introduce the permutation operator $\hat{P}_{ij}$, the effect of which is to exchange the coordinates of particles $i$ and $j$, we can write the two possibilities as follows:

$$\hat{P}_{ij}\Psi = \pm \Psi \qquad (9\text{-}25)$$

If when we exchange the coordinates, the wavefunction remains the same (i.e., $\hat{P}_{ij}\Psi = +\Psi$), we say the function is *symmetric* with respect to the exchange; if the wavefunction changes sign (i.e., $\hat{P}_{ij}\Psi = -\Psi$), we say it is *antisymmetric* with respect to the exchange.

The above discussion is general for indistinguishable particles. We now consider specific indistinguishable particles and postulate that all fundamental particles belong to one or other of the above symmetry types. Furthermore, we postulate that all particles with integral or zero spin are described by symmetric wavefunctions, and all particles with half-integral spin are described by antisymmetric wavefunctions. These postulates are in accord with experimental evidence.

As examples of particles of the first type we can quote: photons (spin = 1), deuterons (spin = 1), and helium nuclei $_2^4\text{He}^{2+}$ (spin = 0), as well as all particles with even mass numbers. Such particles are called *bosons*.

More interesting to us at present are the particles for which the wavefunctions must be antisymmetric. Our interest is due to the fact that electrons (spin = $\frac{1}{2}$) are included in this group. Other examples are protons (spin = $\frac{1}{2}$) and neutrons (spin = $\frac{1}{2}$), and, in general, all particles with odd mass numbers. These particles are referred to as *fermions*.

Let us reiterate what is for our present purposes the most important point in this discussion, i.e., *the wavefunction for a many-electron system is antisymmetric to exchange of the coordinates of any two electrons.* This latter statement is one form of the *Pauli exclusion principle*[5]. Here we have introduced the principle as a postulate, so we should add this postulate to the other basic postulates of quantum mechanics we introduced in Chapter 4.

We can illustrate the application of the Pauli principle by giving further consideration to the helium atom. We shall discuss both the ground state and the first excited state.

We recall that an approximate function for the ground state of the helium atom is given by a product function (cf. Eq. (9-17)). We are now in a position to take into account the spin of the electrons and write the product function in terms of spin-orbitals rather than just space-orbitals. In the ground state both the spatial functions are hydrogen-like $1s$ orbitals, so we can write them as $\phi_{1s}$. Now the spatial functions can be combined with either $\alpha$ or $\beta$ spin functions. Consequently, we can write four simple product functions:

$$\Psi_A = \phi_{1s}(1)\,\alpha(1)\,\phi_{1s}(2)\,\alpha(2)$$

$$\Psi_B = \phi_{1s}(1)\alpha(1)\phi_{1s}(2)\beta(2)$$

$$\Psi_C = \phi_{1s}(1)\beta(1)\phi_{1s}(2)\alpha(2) \qquad (9\text{-}26)$$

$$\Psi_D = \phi_{1s}(1)\beta(1)\phi_{1s}(2)\beta(2)$$

Here the 1 and 2 in parentheses represent the coordinates of electrons 1 and 2.

The four functions in Eq. (9-26) all satisfy the eigenvalue equation resulting from use of the Hamiltonian in Eq. (9-14). However, none of them is acceptable because they do not satisfy the Pauli principle. For example,

$$\hat{P}_{12}\Psi_A = \Psi_A \qquad (9\text{-}27)$$

i.e., $\Psi_A$ is symmetric with respect to exchange of the coordinates of electrons 1 and 2. We can, however, obtain an acceptable function by taking a linear combination of two of the functions given in Eq. (9-26):

$$\frac{1}{\sqrt{2}}[\phi_{1s}(1)\alpha(1)\phi_{1s}(2)\beta(2) - \phi_{1s}(1)\beta(1)\phi_{1s}(2)\alpha(2)] \qquad (9\text{-}28)$$

We can also write this in the following way:

$$\phi_{1s}(1)\phi_{1s}(2)\frac{1}{\sqrt{2}}[\alpha(1)\beta(2) - \beta(1)\alpha(2)] \qquad (9\text{-}29)$$

Equation (9-29) shows the spatial and spin parts of the function separately. The $1/\sqrt{2}$ is a normalization factor. Now Eq. (9-28) (or Eq. (9-29)) is a linear combination of solution functions for the eigenvalue equation we are considering; therefore, it too must be a solution. Furthermore, it is an acceptable function because it is antisymmetric under the permutation operation associated with $\hat{P}_{12}$, a fact you can easily check.

If we examine Eq. (9-29), we note that it is the spin function, i.e., the part in square brackets, that is antisymmetric; the spatial part — the function we used in Section 9-4 to determine the energy of the helium atom — is symmetric. When we examined the energy of the ground state of the helium atom we did not include the spin function. Actually, inclusion of the spin function would not have affected the energy, a fact we can easily prove: Using Eq. (9-29), the value for the energy is given by

$$E = \tfrac{1}{2} \int\!\!\int \phi_{1s}^*(1)\phi_{1s}^*(2)[\alpha(1)\beta(2) - \beta(1)\alpha(2)]^*\hat{H}$$
$$\times \phi_{1s}(1)\phi_{1s}(2)[\alpha(1)\beta(2) - \beta(1)\alpha(2)]\,dv\,d\sigma \qquad (9\text{-}30)$$

where $v$ refers to the spatial coordinates and $\sigma$ to the spin coordinates. Since $\hat{H}$ is a spin-free operator, we can write

$$E = \int \phi_{1s}^*(1)\phi_{1s}^*(2)\hat{H}\phi_{1s}(1)\phi_{1s}(2)\,dv$$
$$\times \int \frac{[\alpha(1)\beta(2) - \beta(1)\alpha(2)]^*[\alpha(1)\beta(2) - \beta(1)\alpha(2)]}{2}\,d\sigma \qquad (9\text{-}31)$$

We now consider the integration over the spin functions; on expansion, this becomes

$$\tfrac{1}{2}\Big[ \int \alpha(1)^*\alpha(1)d\sigma_1 \int \beta(2)^*\beta(2)d\sigma_2 - \int \alpha(1)^*\beta(1)d\sigma_1 \int \beta(2)^*\alpha(2)d\sigma_2$$
$$- \int \alpha(2)^*\beta(2)d\sigma_2 \int \beta(1)^*\alpha(1)d\sigma_1 + \int \alpha(2)^*\alpha(2)d\sigma_2$$
$$\times \int \beta(1)^*\beta(1)d\sigma_1 \Big] = 1 \qquad (9\text{-}32)$$

The last step follows because the spin functions $\alpha$ and $\beta$ are orthogonal and normalized. Thus we see that we are left with the integration over the spatial functions—this was the integral we previously used to determine the energy.

The first excited state of the helium atom is a somewhat more interesting example of the application of the Pauli principle. In the independent-particle model of a many-electron atom—i.e., neglecting the electron-electron repulsion term in the Hamiltonian—the wavefunction for the system is again a product function of hydrogen-like orbitals. However, the spatial functions are now $1s$ and $2s$ orbitals rather than the two $1s$ orbitals we used for the ground state. (We shall discuss the determination of electronic configurations of atoms in the next section; for now we shall just accept that $1s$ and $2s$ orbitals are involved here.) As for the ground state, we can combine each spatial function with either $\alpha$ or $\beta$ spin functions in order to obtain spin-orbitals. Again, the simple product functions we obtain when we use these spin-orbitals are unacceptable because they do not satisfy the Pauli principle. Consequently, we form linear combinations that do satisfy this principle. This time there are four acceptable functions; keeping the spatial and spin functions separate, as in Eq. (9-29), we write them:

$$\frac{1}{\sqrt{2}}[1s(1)2s(2)+2s(1)1s(2)]\,\frac{1}{\sqrt{2}}[\alpha(1)\beta(2)-\beta(1)\alpha(2)]$$

$$\frac{1}{\sqrt{2}}[1s(1)2s(2)-2s(1)1s(2)]\alpha(1)\alpha(2)$$

$$(9\text{-}33)$$

$$\frac{1}{\sqrt{2}}[1s(1)2s(2)-2s(1)1s(2)]\frac{1}{\sqrt{2}}[\alpha(1)\beta(2)+\beta(1)\alpha(2)]$$

$$\frac{1}{\sqrt{2}}[1s(1)2s(2)-2s(1)1s(2)]\beta(1)\beta(2)$$

In the first of these functions the spatial part is symmetric whereas the spin part is antisymmetric, so that the whole function is antisymmetric. In the other three the spatial parts are antisymmetric whereas the spin parts are symmetric—the total functions are therefore antisymmetric. A further point we should note is that the last three functions all have the same spatial functions, differing only in their spin functions. Since the energy depends only on the spatial function, they are degenerate—we say that they form the three components of a *triplet state*. The first function has a different spatial function than the other three—it is called a *singlet state.* The triplet state is an example of *spin degeneracy*, a question we examine in more detail in Section 9-9 when we discuss the total angular momentum of many-electron atoms.

## 9-7  WAVEFUNCTIONS FOR MANY-ELECTRON SYSTEMS: SLATER DETERMINANTS

The examples of the application of the Pauli principle that we have used involve the simple case of a two-electron system. However, we can easily extend the arguments to systems with more than two electrons. To do this, we first write Eqs. (9-28) and (9-29) in an alternate form, because by so doing we can more easily generalize the results we have obtained for the two-electron atom. What we do is express Eqs. (9-28) and (9-29) as a determinant:

$$\frac{1}{\sqrt{2}}\begin{vmatrix} \phi_{1s}(1)\alpha(1) & \phi_{1s}(1)\beta(1) \\ \phi_{1s}(2)\alpha(2) & \phi_{1s}(2)\beta(2) \end{vmatrix} \tag{9-34}$$

This is an example of a *Slater determinant.* We shall now generalize this to show how Slater determinants can be used to express the wavefunctions of many-electron systems.

If we have an $N$-electron system, using the independent-particle model, we can write a product function for it. If there are $N$ spin-orbitals $\psi_i$, one possibility for this function is

$$\Psi' = \psi_1(1)\psi_2(2)\psi_3(3) \cdots \psi_N(N) \tag{9-35}$$

What we have done here is assigned electron 1 to spin-orbital $\psi_1$, electron 2 to spin-orbital $\psi_2$, etc. Since the electrons are indistinguishable, exchange of the coordinates of electrons 1 and 2 amongst the spin-orbitals will give an equally good function:

$$\Psi'' = \psi_1(2)\psi_2(1)\psi_3(3) \cdots \psi_N(N) \tag{9-36}$$

In fact, we can write $N!$ functions of this type—these $N!$ functions would allow for all possible exchanges of the coordinates of the electrons. As in the helium atom case we have just discussed, the simple product functions do not satisfy the Pauli principle. The total wavefunction for the system must be antisymmetric so, following the procedure we used for the helium atom, we write a linear combination of the $N!$ functions that satisfies this condition. This we can easily do by generalizing the form of the Slater determinant of which Eq. (9-34) was a special case. The Slater determinant for an $N$-electron system is

$$\Psi = \frac{1}{\sqrt{N!}}\begin{vmatrix} \psi_1(1) & \psi_2(1) & \cdots & \psi_N(1) \\ \psi_1(2) & \psi_2(2) & \cdots & \psi_N(2) \\ \cdots\cdots\cdots\cdots\cdots\cdots\cdots\cdots\cdots \\ \psi_1(N) & \psi_2(N) & \cdots & \psi_N(N) \end{vmatrix} \tag{9-37}$$

(As before, the $1/\sqrt{N!}$ is a normalization factor.) The linear combination of the spin-orbitals expressed by this determinant is necessarily anti-symmetric. This is a consequence of a well-known property of determinants: If two rows or columns of a determinant are interchanged, the value of the determinant changes sign – interchanging the coordinates of two electrons is equivalent to interchanging two rows of the Slater determinant.

Slater determinants are rather unwieldy when written in full, as in Eq. (9-37). It is therefore normal to use abbreviated forms. One way to abbreviate a Slater determinant is to write only the diagonal elements. We would therefore write Eq. (9-37) as

$$\Psi = |\psi_1(1)\psi_2(2)\cdots\psi_N(N)| \tag{9-38}$$

Using this notation, we would write the ground state wavefunction for the helium atom, Eq. (9-34), as

$$|\phi_{1s}(1)\,\alpha(1)\,\phi_{1s}(2)\,\beta(2)| \tag{9-39}$$

In order to write the excited state wavefunctions given in Eq. (9-33) we need to use linear combinations of Slater determinants; we would write these functions:

$$\frac{1}{\sqrt{2}}\left[|\phi_{1s}(1)\alpha(1)\phi_{2s}(2)\beta(2)| - |\phi_{1s}(1)\beta(1)\phi_{2s}(2)\alpha(2)|\right]$$

$$|\phi_{1s}(1)\alpha(1)\phi_{2s}(2)\alpha(2)|$$

$$\tag{9-40}$$

$$\frac{1}{\sqrt{2}}\left[|\phi_{1s}(1)\alpha(1)\phi_{2s}(2)\beta(2)| + |\phi_{1s}(1)\beta(1)\phi_{2s}(2)\alpha(2)|\right]$$

$$|\phi_{1s}(1)\beta(1)\phi_{2s}(2)\beta(2)|$$

Another way of abbreviating Slater determinants emphasizes the anti-symmetric property by using an *antisymmetrization operator* $\hat{A}$ – we write for Eq. (9-37):

$$\Psi = \hat{A}\psi_1(1)\psi_2(2)\cdots\psi_N(N) \tag{9-41}$$

## 9-8   PERIODIC SYSTEM OF THE ELEMENTS

The periodic system of the elements and the electronic configurations of the atoms that are associated with the periodic table will be familiar to you from introductory chemistry courses. Here we shall very briefly review this subject in the light of what we have been discussing in this chapter.

In the approximation where we use product functions for a many-electron system, we write the total wavefunction as a Slater determinant (cf. Eq. (9-37)). Now a property of determinants is that if two columns are identical, the value of the determinant is zero. The significance of this is that no spin-orbital can be occupied by more than one electron. If, for example, two electrons were in a single spin-orbital, say $\psi_1$, the Slater determinant would be

$$\begin{vmatrix} \psi_1(1) & \psi_1(1) & \psi_2(1) & \cdots & \psi_N(1) \\ \psi_1(2) & \psi_1(2) & \psi_2(2) & \cdots & \psi_N(2) \\ \cdots\cdots\cdots\cdots\cdots\cdots\cdots\cdots\cdots\cdots\cdots \\ \psi_1(N) & \psi_1(N) & \psi_2(N) & \cdots & \psi_N(N) \end{vmatrix} = 0 \qquad (9\text{-}42)$$

Now the orbitals we are discussing here are hydrogen-like in form — we can label them with the four quantum numbers $n$, $l$, $m$, and $m_s$. We can then express what we have just said as follows: *No two electrons can have the same set of four quantum numbers.* This latter statement is an alternative and important form of the Pauli principle.

To determine the ground state of a many-electron atom we have to assign the electrons so that the spin-orbitals are singly occupied, filling these orbitals in order of increasing energy. When we make these assignments in this way we are using what is known as the *aufbau (or building-up) principle.* The result is the *electronic configuration* of the atom. Thus we see that the Pauli principle plays a very important part in the explanation of the configurations of the atoms.

There is one further point we must discuss before we examine the actual configurations for the elements — this is the relative energies of the orbitals. The space-orbitals we are working with are hydrogen-like. We therefore label them $1s$, $2s$, $2p$, $3s$, $3p$, $3d$, etc. Now in the hydrogen atom the order of energies for these orbitals is:

$$1s < 2s = 2p < 3s = 3p = 3d < 4s = 4p = 4d = 4f \cdots$$

However, we must remember that this order refers to a single-electron system where there are, of course, no electron-electron interactions. In a many-electron atom, one effect of electron-electron interactions is to remove the degeneracies associated with the quantum number $l$. We can explain this in terms of our approximation where we used effective nuclear charges in order to partially account for electron-electron interaction. Using the orbitals with $n = 2$ as an example, we can say that an electron in a $2s$ orbital will experience a different screening by the $1s$ electrons than an electron in a $2p$ orbital—this is due to the different electron probability distributions for $2s$ and $2p$ orbitals (cf. Fig. 6-7). The maximum in the probability distribution for a $2s$ electron is nearer the nucleus than the maximum for a $2p$ electron. This results in the $2s$ electron penetrating the $1s$ shell to a greater extent than the $2p$ electron. Therefore, the effective nuclear charge for a $2s$ electron is greater than for a $2p$ electron. Consequently, a $2p$ orbital is higher in energy than a $2s$ orbital.

On the basis of arguments such as these we can construct the familiar energy level diagrams that are commonly used to picture the electronic configurations of atoms. Before we actually look at these configurations, we should note that $p, d, f$, etc. orbitals are degenerate because the quantum number $m$ can take several values; thus, for example, there are three $p$ orbitals corresponding to $m = +1, 0, -1$.

We are now in a position to examine the ground state electronic configurations of the elements. To determine the ground state of hydrogen we assign the single electron to the orbital of lowest energy, i.e., the $1s$ orbital. There are two spin-orbitals corresponding to the $1s$ space-orbital, so for helium both electrons can be assigned to the $1s$ orbital—we write the configuration $(1s)^2$. The first shell, called the $K$ shell, is now complete, so helium is an inert or noble gas. The next element, lithium, has one electron in the $2s$ orbital, so its configuration is written $(1s)^2(2s)$ or, more concisely, $K(2s)$. As for the $1s$ orbital, there are two spin-orbitals associated with the $2s$ space-orbital; therefore, the next element to lithium, beryllium, has the electronic configuration $K(2s)^2$. And so on, until we reach neon $K(2s)^2(2p)^6$—this completes the shell with $n = 2$, i.e., the $L$ shell, making neon an inert gas like helium. We continue to fill the $3s$ and $3p$ orbitals to arrive at argon $KL(3s)^2(3p)^6$. $4s$ electrons penetrate the inner shells to a greater extent than $3d$ electrons do; consequently, the $4s$ orbital is lower in energy than the $3d$ orbital—the next element to argon, potassium, has an electron in the $4s$ orbital to give a configuration $KL(3s)^2(3p)^6(4s)$. After calcium $KL(3s)^2(3p)^6(4s)^2$ the $3d$ orbitals are filled to give a series of transition elements, starting with scandium $KL(3s)^2(3p)^6(3d)(4s)^2$.

We shall not continue this discussion since you will be very familiar with the electronic configurations of the elements and the relationship to the positions of the elements in the periodic table. For a detailed discussion you can consult any introductory physical chemistry book. We should, however, just mention that the level of approximation we have used here is not good enough to predict the electronic configurations of some elements, e.g., chromium and copper. For example, on the basis of the above arguments, we would expect chromium to have the configuration $KL(3s)^2(3p)^6(3d)^4(4s)^2$. In fact, it has the configuration $KL(3s)^2$-$(3p)^6(3d)^5(4s)$. We shall not go into the details of explaining these apparent anomalies.

## 9-9   ANGULAR MOMENTUM OF MANY-ELECTRON ATOMS

We should now have some idea about how to deal with the total energy and wavefunctions of many-electron atoms. What about their angular momenta?

We saw in Chapter 5 that each electron in an atom has associated with it an orbital angular momentum and a spin angular momentum. Presumably these can be combined to give a total angular momentum if we have a many-electron atom. The question is: How do we do this? There are two commonly used methods—we call them *Russell-Saunders, or LS, coupling* and *jj coupling*. In the Russell-Saunders scheme the individual electron orbital angular momenta are first combined to give a resultant, then the individual electron spin angular momenta are combined to give a second resultant; finally, these two resultant angular momenta are combined to give the total angular momentum. Since angular momenta can be represented by vectors, the method of combination will be vector addition. In *jj* coupling the orbital and spin angular momenta for each electron are first combined; then all the resultant vectors are combined. Russell-Saunders coupling is more appropriate for light atoms; *jj* coupling is used for heavy atoms and highly excited states of light atoms. Actually both are approximations, the true picture being something between the two. Here, however, we shall not deal with *jj* coupling any further since Russell-Saunders coupling is the important method for most atoms of practical chemical interest. We now consider Russell-Saunders coupling in more detail.

Let the orbital angular momenta for the electrons be $\mathbf{l}_i$; let the spin angular momenta be $\mathbf{s}_i$. In the Russell-Saunders method we first combine these as follows to give total orbital, **L**, and total spin, **S**, angular momenta:

$$\mathbf{L} = \sum_i \mathbf{l}_i \qquad \mathbf{S} = \sum_i \mathbf{s}_i \tag{9-43}$$

Then we combine **L** and **S** to get a total angular momentum **J**;

$$\mathbf{J} = \mathbf{L} + \mathbf{S} \tag{9-44}$$

Now there will be operators associated with **J**, **L**, and **S**, and also with the squares and components of the angular momenta. In particular, we can define the operators $\hat{L}^2$, $\hat{L}_z$, $\hat{S}^2$, $\hat{S}_z$, $\hat{J}^2$, and $\hat{J}_z$ associated with $L^2$, $L_z$, $S^2$, $S_z$, $J^2$, and $J_z$. We are particularly interested in these because they commute with the Hamiltonian (provided we do not include in the Hamiltonian, terms arising from the coupling between spin and orbital motions of the electrons). This means that eigenstates of the Hamiltonian are also eigenstates of these other operators. Therefore, the quantum numbers associated with these operators can be used to label the energy states of the atom. We shall call these quantum numbers $L$, $M_L$, $S$, $M_S$, $J$, and $M_J$ respectively.

How do we determine the values of the complete set of quantum numbers associated with the total angular momentum? We recall from Section 5-3 that the magnitude of an angular momentum vector **M** is given by $[l(l+1)]^{1/2}\hbar$ where $l$ is the quantum number associated with $M^2$. For two combining angular momentum vectors, for example, $\mathbf{l}_1$ and $\mathbf{l}_2$, we get the resultant **L**; the magnitude of **L** is $[L(L+1)]^{1/2}\hbar$. $L$ takes the values $l_1 + l_2, l_1 + l_2 - 1, \ldots, l_1 - l_2$ for $l_1 \geqslant l_2$. $M_L$ then takes the $2L+1$ values $L$, $L-1, \ldots, -L$. Similarly, when two spin angular momentum vectors $\mathbf{s}_1$ and $\mathbf{s}_2$ combine $S$ takes the values $s_1 + s_2, s_1 + s_2 - 1, \ldots, s_1 - s_2$, and $M_s$ takes the $2S+1$ values $S, S-1, \ldots, -S$. When we combine **L** and **S** as in Eq. (9-44) to get **J**, the quantum number $J$ associated with the total angular momentum takes the values $L+S, L+S-1, \ldots, L-S$ for $L \geqslant S$, or $S+L, S+L-1, \ldots, S-L$ for $S > L$. Figure 9-2 is a diagrammatic representation of this latter case for $L = 2$ and $\mathbf{S} = \frac{3}{2}$.

We usually summarize the angular momentum state of an atom by a *term symbol*. We write this symbol $^{2S+1}L_J$. For example, if you see the term symbol $^3F_2$ for a state of an atom, this means that $L = 3$, $S = 1$ and $J = 2$. In general we use the letters $S, P, D, F, G, \ldots$ to label states with $L = 0, 1, 2, 3, 4, \ldots$ respectively. The *multiplicity* of the state is the superscript $2S + 1$. The $^3F$ term we used as an example has three *levels* associated with it; these are labeled $^3F_4$, $^3F_3$, $^3F_2$, showing the different values for $J$. $J$, of course, can take half-integral values since $S$ can take half-

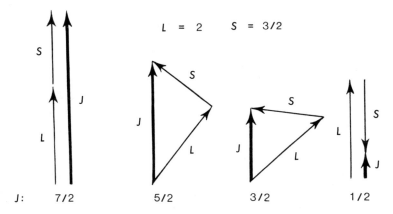

**Fig. 9-2**    Vector addition for $L = 2$ and $S = \frac{3}{2}$.

integral values. You will, therefore, in your reading see term symbols such as $^2P_{1/2}$, $^4S_{3/2}$, and so on.

Let us consider some examples. First we should note that any atom that has a closed-shell configuration (e.g., $p^6$ or $f^{14}$) gives only a $^1S$ term. Why is this so? Remember that $M_L$ refers to the component of the orbital angular momentum in a specified direction. $M_L$ for a closed-shell configuration is zero, because each positive value of $m$ for the individual electrons is cancelled by its negative value. Since zero is the only value for $M_L$, $L$ must be zero also; i.e., we get an $S$ term. The Pauli exclusion principle requires that every electron has its spin angular momentum paired; therefore, $S$ is zero also, and $2S + 1 = 1$; i.e., we have a singlet state.

Let us now consider an atom with all filled shells except two different $p$ shells each with one electron in it. The closed shells will not contribute anything to the total angular momentum. Therefore, we need only consider the two $p$ electrons. For these $l_1 = 1$ and $l_2 = 1$; therefore, $L$ takes the values 2, 1, 0. Considering the spin angular momentum, $s_1 = \frac{1}{2}$ and $s_2 = \frac{1}{2}$; therefore, $S$ takes the values 1 or 0. The possible terms are, therefore, $^1D, \,^3D, \,^1P, \,^3P, \,^1S, \,^3S$.

The latter example involved just simple vector addition of the angular momenta. This works so long as we do not have more than one electron in any of the unfilled sets of orbitals $s, p, d, f$, etc. If there is more than one electron in an unfilled shell, we have to make sure that the Pauli exclusion principle is obeyed when determining which terms actually arise in practice. We shall not do this; if you are interested in the details, you can consult one of the references in the Bibliography (e.g., reference B6).

## 9-10 RELATIVE ENERGIES OF ATOMIC STATES

We have said nothing yet about the relative energies of the different terms we obtain when we determine the total angular momentum of the electrons in an atom. In fact we shall not go into the details of this subject, but merely indicate some of the factors involved. If we have several terms for an atomic ground state configuration, we can distinguish the term corresponding to the lowest energy by using some empirical rules called *Hund's rules*. These rules are: 1. For a given configuration the term of lowest energy is the one with greatest multiplicity. 2. If two terms have the same multiplicity, the term corresponding to the largest angular momentum has the lowest energy. For configurations of higher energy than the ground state, there are frequent exceptions to these rules, though they may still be used as a guide.

At this point we shall just mention that there is something that affects the energies associated with terms that we have not taken into account. This is *spin-orbit coupling*. Spin-orbit coupling is due to interaction between the spin magnetic moment of the electrons and the magnetic field set up by the motion of the electrons about the nucleus. Though we shall not do it here, we can deal with spin-orbit coupling by adding a term to the Hamiltonian of Eq. (9-5). The effect of spin-orbit coupling is to resolve the degeneracy in states that differ only in their $J$ values. For example, atomic nitrogen has a $^2P$ term; spin-orbit coupling resolves this into two terms $^2P_{1/2}$ and $^2P_{3/2}$. The effect of spin-orbit coupling on the nitrogen atom is shown in Fig. 9-3.

We should also mention that the states we obtain by taking account of spin-orbit coupling may still be degenerate. We can resolve this degeneracy by applying a magnetic field to the atom when states with different values of $M_J$ are split. In fact, there will be $2J + 1$ states corresponding to a particular value of $J$; each has a different value for $M_J$, the quantum number associated with the component of the total angular momentum in a particular direction. This further splitting of states in a magnetic field we call the *Zeeman effect*.

## 9-11 HARTREE-FOCK SELF-CONSISTENT FIELD METHOD

When trying to determine the energies and wavefunctions for many-electron atoms, we have attempted to account for the effects of the electrons on each other by using effective nuclear charges. The wavefunctions were hydrogen-like orbitals with the effective nuclear charges

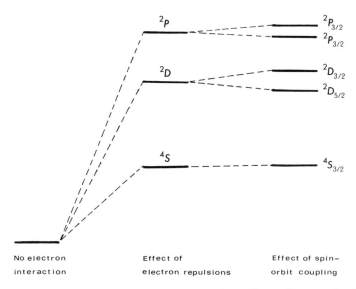

**Fig. 9-3**   Effects of electron repulsion and spin-orbit coupling on the energy levels of atomic nitrogen. (Not to scale.)

substituted for the actual nuclear charges. In the case of the ground state of the helium atom this approximation led to an energy that differed from the experimental energy by 0.057 hartrees. This is about 35 kcal mole$^{-1}$ which is not inconsiderable in experimental terms. Again, therefore, we would like to improve the wavefunctions for many-electron atoms.

If we review the best procedure we have so far used, we realize that the variation method has been carried through with only one variation parameter (the effective nuclear charge) per orbital. We might therefore expect that we could improve the wavefunctions by a more systematic application of the variation theorem. The procedure whereby we can do this is called the *Hartree-Fock SCF* (self-consistent field) method.

The method was first developed in the early 1930's by Hartree[6], Fock[7] and Slater[8], and was originally applied to atoms with closed shells or with one electron or one hole in a shell. It has since been extended to open-shell systems. Here we shall limit our discussion to the closed-shell case; the wavefunction for this type of system can be suitably expressed as a single Slater determinant (cf. Eq. (9-34)). For the open-shell case, a single Slater determinant is not necessarily a good function (cf. Eq. (9-40)). We shall not go into the details of the method since these are considerably beyond the scope of this book, but shall just very briefly mention the major points†.

Consider the ground state of an $N$-electron atom for which the wavefunction $\Psi$ can be written as a single antisymmetrized product of spin-orbitals $\psi_i$ (cf. Eq. (9-37)):

$$\Psi = |\psi_1(1)\psi_2(2) \cdots \psi_N(N)| \qquad (9\text{-}45)$$

According to the variation theorem, if this is an arbitrarily chosen function of this form, we can use it as a trial function and minimize the energy of the system in order to obtain the best function of the type; specifically,

$$\frac{\int \Psi^* \hat{H} \Psi d\tau}{\int \Psi^* \Psi d\tau} \geq E_0$$

where $E_0$ is the lowest eigenvalue of the Hamiltonian. If we can effect a complete variation of the $\psi_i$, we will get the best wavefunction possible using a single determinant. The spin-orbitals we obtain are then the Hartree-Fock orbitals and they are the best set of one-electron wavefunctions obtainable with a function of the type Eq. (9-45).

When we make a complete variation of the spin-orbitals in Eq. (9-45) the many-electron problem leads to a set of coupled one-electron equations, the Hartree-Fock equations:

$$\begin{aligned}
\hat{H}_{\text{eff}}(1)\psi_1(1) &= E_1\psi_1(1) \\
\hat{H}_{\text{eff}}(2)\psi_2(2) &= E_2\psi_2(2) \\
&\cdots\cdots\cdots\cdots \\
\hat{H}_{\text{eff}}(N)\psi_N(N) &= E_N\psi_N(N)
\end{aligned} \qquad (9\text{-}46)$$

Here $\hat{H}_{\text{eff}}$ is the effective one-electron Hamiltonian, and it is the same for each electron; $E_i$ are the *orbital energies*.

We now examine the form of $\hat{H}_{\text{eff}}$. This, as usual, is made up of a kinetic energy term and a potential energy term. The problem is to determine the form of the latter term. To do this we assume that each electron is subject to the potential of the nucleus plus some average, spherically symmetric potential field created by all the other electrons. Because of this spherical symmetry, the angular parts of the functions are the familiar spherical harmonics that we encountered in the hydrogen atom problem. With this

†For a much more detailed discussion you can consult reference C2 in the Bibliography pp. 186–193 and Chapter 13.

assumption we can then write $\hat{H}_{\text{eff}}$ as

$$\hat{H}_{\text{eff}}(i) = \hat{h}(i) + \hat{U}_{\text{av}}(i) \qquad (9\text{-}47)$$

Here $\hat{h}(i)$ is a hydrogen-like Hamiltonian for electron $i$; $\hat{U}_{\text{av}}(i)$ accounts for the average field of the other electrons. We shall not go into the details of determining $\hat{U}_{\text{av}}$, but we can note that it has the form

$$\hat{U}_{\text{av}}(i) = \sum_{j=1}^{N} \int \psi_j^*(j) \left[ \frac{(1 - \hat{P}_{ij})}{r_{ij}} \right] \psi_j(j) d\tau_j \qquad (9\text{-}48)$$

where the integration is over the coordinates of the $j$th electron. The permutation operator $\hat{P}_{ij}$ is a result of our requiring an antisymmetrized wavefunction.

We have given the form of the operator $\hat{U}_{\text{av}}$ to emphasize that the Hamiltonian for use in Eq. (9-46) contains the orbitals $\psi_i$ which, of course, are initially unknown. We therefore have to solve the Hartree-Fock equations by an iterative method. We start by making a guess at the form of the $\psi_i$'s. Then we use this set of functions to determine the potential acting on one electron, say electron $k$. This allows us to solve for electron $k$'s orbital. We then repeat the procedure for another electron's orbital, ending up with another improved orbital. The process is repeated until all the original orbitals have been replaced by improved orbitals. We then take the improved set of orbitals and start the whole procedure again in order to get a second improved set of orbitals. Continuing in this way through third, fourth, etc. improved sets of orbitals, we eventually obtain a set of orbitals that does not change (within the desired accuracy) after an iteration. At this point we say we have obtained a self-consistent field because the final set of orbitals is able to reproduce the potential field that determined them.

To conclude this brief examination of the Hartree-Fock SCF method we should make a few comments about the total electronic energy $E_{\text{HF}}$ calculated using the orbitals defined by the procedure we have just described. If we were to work out an expression for this energy, we would discover it to be given by

$$E_{\text{HF}} = \sum_{i=1}^{N} E_i - \sum_{i>j}^{N} (J_{ij} - K_{ij}) \qquad (9\text{-}49)$$

The terms in this equation have the following significance: $E_i$ are the orbital energies, i.e.,

$$E_i = \int \psi_i^* \hat{H}_{\text{eff}} \psi_i d\tau \qquad (9\text{-}50)$$

$J_{ij}$ is an abbreviated notation for the integrals given by

$$J_{ij} = \int \psi_i^*(1)\psi_j^*(2) \left(\frac{1}{r_{12}}\right) \psi_i(1)\psi_j(2) d\tau_1 d\tau_2 \qquad (9\text{-}51)$$

These integrals are called *coulomb integrals*. $J_{ij}$ represents the interaction between the electron densities $|\psi_i|^2$ and $|\psi_j|^2$. $K_{ij}$ is given by

$$K_{ij} = \int \psi_i^*(1)\psi_j^*(2) \left(\frac{1}{r_{12}}\right) \psi_i(2)\psi_j(1) d\tau_1 d\tau_2 \qquad (9\text{-}52)$$

These integrals arise because of the permutation operator in $\hat{H}_{\text{eff}}$; i.e., they are a result of the exchange symmetry of the electrons – they are called *exchange integrals*.

### 9-12  CORRELATION ENERGY

Energies calculated by the Hartree-Fock method are usually within about one percent of the experimental energies. This is quite good in absolute terms, but for most chemical purposes these results are not usually very satisfactory. In chemical problems we are generally interested in energy differences, e.g., the energy difference between two spectroscopic states, or between the reactants and products of a chemical reaction. Small errors in the absolute energies of states of systems can therefore lead to large relative errors in energies of practical interest.

If we further examine the wavefunctions obtained from Hartree-Fock calculations, we can immediately understand why there is a residual error in the energy calculations. These wavefunctions result from taking antisymmetrized product functions to describe our systems. The problem with this type of function is that it does not fully correlate the motions of the electrons. What we mean by this is that the function assumes that each electron's motion is to some extent independent of the positions of the other electrons. Electrons with the same spin are not allowed in the same place at the same time due to the antisymmetric nature of the function, but the function does not satisfactorily keep electrons of opposite spin from being in the same region of space. When electrons come close together there is a very strong repulsive effect, so we can expect that the Hartree-Fock energy will be in error due to a positive contribution resulting from lack of proper electron correlation. We therefore define the correlation energy of a state of a system $E_{\text{corr}}$ as

$$E_{\text{corr}} = E - E_{\text{HF}}$$

Here $E$ is the "exact" energy for the specified Hamiltonian for the system; i.e., it is the exact eigenvalue of the specified Hamiltonian. $E_{HF}$ is the energy calculated from the Hartree-Fock approximation. Here we have been considering non-relativistic Hamiltonians, so we define the non-relativistic correlation energy. It is important to realize that this latter correlation energy is not the difference between the exact experimental energy and the Hartree-Fock energy — the experimental energy contains relativistic contributions.

## 9-13  SLATER-TYPE ORBITALS

The Hartree-Fock method gives orbital functions that are rather unwieldy. It is common therefore to approximate these functions by what are known as *Slater-type orbitals (STO's)*[9]. STO's have the general form

$$R_{n,l}(r)Y_{l,m}(\theta,\phi) \qquad (9\text{-}53)$$

where the radial part is given by

$$R(r) = Nr^{(n^*-1)}e^{-\zeta r} \qquad (9\text{-}54)$$

$N$ is a normalizing constant; $n^*$ is known as the *effective quantum number*; $\zeta$ is a parameter that allows for the screening of the nucleus by the electrons. Slater has given some simple rules whereby we can determine satisfactory values for $n^*$ and $\zeta$ — they are:

1. $n^*$ is related to $n$, the principal quantum number in the following way:

$$
\begin{array}{lcccccc}
n\text{:} & 1 & 2 & 3 & 4 & 5 & 6 \\
n^*\text{:} & 1 & 2 & 3 & 3.7 & 4.0 & 4.2
\end{array}
$$

2. The parameter $\zeta$ is given by $\zeta = (Z - S)/n^*$. $Z$ is the nuclear charge; $S$ is called the *screening constant* because it takes account of the screening of the nucleus by the electrons.

3. We calculate $S$ by considering the screening by each group of electrons. For this purpose the electrons are classified in the following groups of orbitals:

$$(1s), (2s\ 2p), (3s\ 3p), (3d), (4s\ 4p), (4d), (4f), \text{etc.}$$

4. $S$ is calculated by summing the following contributions:

(a)  Nothing from electrons in groups higher than the one in question.

(b)  Every electron in the same group contributes 0.35 (except for the $1s$ group which gives 0.30).

(c) For an *sp* group each electron in the next lower group gives 0.85. If the group is *d*, *f*, or *g* each inner electron contributes 1.00.

(d) For groups lower by two or more from the one containing the electron under consideration, each electron contributes 1.00.

As an example of the use of Slater's rules we can consider a 3*s* electron in the sulfur atom $(1s^2 2s^2 2p^6 3s^2 3p^4)$. For this electron the screening constant is given by

$$S = (5 \times 0.35) + (8 \times 0.85) + (2 \times 1) = 10.55$$

This means that the effective charge experienced by the electron is not the actual charge 16, but $16 - 10.55 = 5.45$.

It is interesting to compare the result of using Slater's rules for the ground state of the helium atom with the result calculated in Section 9-4 where we used the variation theorem. Slater's rules give a value of 1.70 for the effective nuclear charge. This compares quite well with the value of 1.69 given by the variation procedure.

## SUMMARY

1. We examined the non-relativistic Hamiltonian for a many-electron atom and discovered that it is the term associated with the electron-electron repulsions that causes the mathematical difficulty when we try to solve the Schroedinger equation for a many-electron atom.

2. We examined an approximation method for solving the Schroedinger equation—this involved omitting the electron-electron repulsion term from the Hamiltonian, the result being the reduction of the equation to a series of hydrogen-like equations.

3. We applied this approximation method to the simplest many-electron atom, the helium atom. This gave the ground state wavefunction and energy. The result, however, was not very good, so we improved the approximation by introducing an *effective nuclear charge* as a variation parameter, and applied the variation theorem to determine the best value for the effective nuclear charge.

4. Next we examined electron spin and total wavefunctions for many-electron systems. The important point to emerge was that total wave-functions for many-electron systems must be antisymmetric to exchange of the coordinates of any pair of electrons (the *Pauli exclusion principle*). This can be accomplished by expressing these wavefunctions as *Slater determinants*. Slater determinants are linear combinations of *spin-*

*orbitals.* (A spin-orbital is a function that combines the spatial and spin functions for an electron.)

5. We noted that the Pauli principle plays a very important part in determining the electronic configurations of the atoms in the periodic table.

6. We considered how to combine the orbital and spin angular momenta for a many-electron atom. In the *Russell-Saunders method*, which is the most important, all the orbital momenta $l_i$ are first combined, then the spin momenta $s_i$ are combined; finally the total orbital $L$ and total spin $S$ momenta are combined to give the total angular momentum $J$.

7. We examined the quantum numbers associated with the angular momentum operators, and summarized the angular momentum of an atom by a *term symbol* which we write as $^{2S+1}L_J$; the superscript is the *multiplicity* of the state.

8. We very briefly discussed the *Hartree-Fock self-consistent field* method for constructing atomic orbitals for atoms with closed-shell configurations. To use this method an initial set of orbital functions is guessed. These are then improved by use of the variation theorem and an iterative method whereby a self-consistent field is achieved.

9. Even after performing a Hartree-Fock SCF calculation there is a residual error in the energy of the many-electron system due to incomplete correlation of the electrons' motions. We defined *correlation energy* as the difference between the Hartree-Fock energy and the exact eigenvalue of the specified Hamiltonian for the system.

10. We concluded the chapter by mentioning that the orbital function for a many-electron atom can be approximated by a *Slater-type orbital (STO)* — the angular part of this is a spherical harmonic; the radial part is given by

$$R(r) = Nr^{(n^*-1)} e^{-\zeta r} \tag{9-54}$$

$n^*$ and $\zeta$ (which accounts for screening of the electrons) can be determined by *Slater's rules.*

## EXERCISES

**9-1**  Convert the Hamiltonian in Eq. (9-1) to atomic units and thereby show that Eq. (9-5) is correct.

**9-2**  Show that the application of the separation of variables technique to Eq. (9-8) leads to the set of equations in Eq. (9-10).

**9-3**  Using the permutation operator notation we introduced at the beginning of Section 9-6, show that the functions in Eq. (9-26) do not satisfy the Pauli principle, but that the function in Eq. (9-28) does.

**9-4**   Show that the normalization factor $(1/\sqrt{2})$ in Eqs. (9-28) and (9-29) is correct.

**9-5**   Use the Laplace development of a determinant to expand the Slater determinant given by

$$\Psi = \hat{A}\psi_1(1)\psi_2(2)\psi_3(3)$$

and show that it is an eigenfunction of the operator $\hat{P}_{13}$ with the eigenvalue $-1$; i.e., confirm for this simple case that interchange of the first and third rows of the determinant results in a change in the sign of $\Psi$.

**9-6**   What will be the multiplicities of the states that arise from three electrons in different orbitals?

**9-7**   What terms arise from the atomic configurations $1s^2 2s^1$ and $1s^2 2s^2 2p^2$? (Hint: In the second case consider the possible ways of placing the electrons and use the Pauli principle to determine which of these possible ways are acceptable.)

**9-8**   Use Slater's rules to calculate the effective nuclear charges for the $K$ and $L$ electrons of the ground state of the carbon atom. Compare the results with the actual nuclear charge.

# 10

# Molecular-orbital Theory

## 10-1  INTRODUCTION

So far in this book we have considered only very simple, idealized systems like the rigid rotator and the harmonic oscillator, and the very simplest of systems of real chemical interest, the hydrogen and helium atoms. We had to start at this level of simplicity in order to present the fundamental ideas of quantum mechanics, and also to indicate how to set about getting exact solutions to quantum mechanical problems. But the real use of quantum mechanics to chemists is in the help it can give us when we try to understand molecular properties. We are especially interested in the nature of chemical bonding; we are also interested in explaining theoretically experimental results such as spectroscopic measurements, dipole moments and ionization potentials. In addition, we would like to be able to make calculations of molecular properties that are experimentally inaccessible – the molecules might have too transient an existence, or the conditions of temperature and pressure make it impossible to get reliable experimental results. In a book of this type we cannot cover all the applications of quantum mechanics in chemistry. We shall, however, treat some of the most important; in particular we shall examine the nature of molecular bonding, and most of the rest of the book is devoted to this question.

Unfortunately, molecular systems are generally far too complex for us to attempt to obtain exact solutions to their quantum mechanical equations. Though computers now allow us to do molecular calculations that used to be impossibly lengthy, we still have to resort to many approximations. We expect, therefore, that the results may not always be in very good agreement with experiment. Nevertheless, they certainly encourage us to believe that quantum mechanics gives a true description of molecular systems.

In this chapter we consider one of the major methods for dealing with chemical bonding—*molecular-orbital (MO) theory*; in the next chapter we consider the other major method—*valence-bond theory*. Before we actually examine molecular-orbital theory, we consider what is known as the Born-Oppenheimer approximation; this allows us to consider the nuclei as having fixed positions with regard to treatment of the electronic wavefunctions. We start our examination of MO theory by looking at the usual method for constructing molecular orbitals, the linear combination of atomic orbitals method. We apply the method to the simplest molecule, the $H_2^+$ ion. Then we briefly examine systems that are somewhat more complex—the hydrogen molecule and other diatomic molecules.

## 10-2 THE BORN-OPPENHEIMER APPROXIMATION

We start this examination of molecular systems by considering an approximation that greatly facilitates our treatment of molecular wavefunctions. This is the *Born-Oppenheimer approximation*[10]. The approximation allows us to assume that the nuclei have fixed positions so far as the study of the electronic wavefunctions is concerned. The physical picture that justifies the Born-Oppenheimer approximation is one where the nuclei move very slowly relative to the electronic motions. Consequently, we are able to specify the electronic wavefunctions quite accurately for the various fixed positions of the nuclei. This slow relative motion of the nuclei is a result of the nuclear masses being thousands of times greater than the electronic mass. Born and Oppenheimer, in their original treatment, expanded the solutions of the Schroedinger equation for the wavefunctions of a molecular system in a power series in $M^{-1/4}$, where $M$ is the average nuclear mass in atomic units. Provided $M$ is very small—and we have just noted that it is—the solutions can then be approximated in a way that we shall now describe. We shall not go into the details of the validity of the approximation, but will just briefly examine what it means with regard to the Hamiltonian for the molecular system and the wavefunctions associated with the Hamiltonian.

The spin-free, non-relativistic Hamiltonian for a molecule with nuclei labeled by $\mu$ and $\nu$ and electrons labeled by $i$ and $j$ is (in atomic units)

$$\hat{H} = -\frac{1}{2}\sum_\mu \frac{\nabla_\mu^2}{M_\mu} - \frac{1}{2}\sum_i \nabla_i^2 - \sum_i \sum_\mu \frac{Z_\mu}{r_{i\mu}} + \sum_{\mu<\nu} \frac{Z_\mu Z_\nu}{R_{\mu\nu}} + \sum_{i<j}\frac{1}{r_{ij}} \quad (10\text{-}1)$$

The first term refers to the kinetic energy of the nuclei ($M_\mu$ is the mass of the $\mu$th nucleus). The second refers to the kinetic energy of the electrons. The third takes account of nucleus-electron attraction potential energies ($r_{i\mu}$ is the distance between the $i$th electron and the $\mu$th nucleus). The fourth takes account of nucleus-nucleus repulsion energies ($R_{\mu\nu}$ is the distance between the $\mu$th and $\nu$th nuclei). Finally, the last term accounts for the electron-electron repulsion energies ($r_{ij}$ is the distance between the $i$th and $j$th electrons).

The Hamiltonian of Eq. (10-1) satisfies the Schroedinger equation

$$\hat{H}\Psi(r,R) = E\Psi(r,R) \qquad (10\text{-}2)$$

Here $E$ is the total energy of the system except for the translational energy of the molecule. $\Psi$ is a function of the electronic coordinates $r$ and the nuclear coordinates $R$. The Born-Oppenheimer approximation allows us to approximate $\Psi$ by writing it as a product function:

$$\Psi(r,R) = \Psi_R(r)\chi(R) \qquad (10\text{-}3)$$

$\Psi_R(r)$ is called the *electronic wavefunction*—for fixed nuclei it is a function of the electronic coordinates $r$ and depends only on the quantum states of the electrons; however, this function varies for different fixed nuclear positions, and it is for this reason that we have indicated with the subscript $R$ the parametric dependence on the nuclear coordinates. $\Psi_R(r)$ satisfies the following equation:

$$\hat{H}_e(r,R)\Psi_R(r) = E_e(R)\Psi_R(r) \qquad (10\text{-}4)$$

$\hat{H}_e$ is called the *electronic Hamiltonian*—it is the Hamiltonian in Eq. (10-1) with the nuclear kinetic energy term removed. The energy eigenvalue $E_e(R)$ is, like the wavefunction, parametrically dependent on the nuclear coordinates. $\chi(R)$ is called the *nuclear wavefunction*—it is a function of the nuclear coordinates but depends on the electronic energy. In fact, the electronic energy can be taken as the potential in which the nuclei move, so that the equation for $\chi(R)$ is:

$$[\hat{H}_n(R) + E_e(R)]\chi(R) = E\chi(R) \qquad (10\text{-}5)$$

Here $\hat{H}_n$ is the first term in the Hamiltonian of Eq. (10-1), i.e., the nuclear kinetic energy term.

Equations (10-3), (10-4), and (10-5) are the equations that describe and are a result of the Born-Oppenheimer approximation. We shall now very briefly examine their validity. Substitute Eq. (10-3) in Eq. (10-2); we get

$$[\hat{H}_n(R) + \hat{H}_e(r,R)]\Psi_R(r)\chi(R) = E\Psi_R(r)\chi(R) \qquad (10\text{-}6)$$

where we have split the Hamiltonian $\hat{H}$ into its two parts, as we did earlier. We now examine in more detail the operations associated with these parts:

$$\hat{H}_n(R)\,[\Psi_R(r)\chi(R)] = -\frac{1}{2}\sum_\mu \frac{1}{M_\mu}[\Psi_R(r)\nabla_\mu^2\chi(R)$$

$$+ 2\nabla_\mu\Psi_R(r)\nabla_\mu\chi(R) + \chi(R)\nabla_\mu^2\Psi_R(r)] \qquad (10\text{-}7)$$

$$\hat{H}_e(r,R)\,[\Psi_R(r)\chi(R)] = \chi(R)\hat{H}_e(r,R)\Psi_R(r) = \chi(R)E_e(R)\Psi_R(r) \qquad (10\text{-}8)$$

We have been able to write the first equality in Eq. (10-8) because the only differential term in $\hat{H}_e$ is a function only of $r$, and not of $R$; $\chi(R)$, of course, is not a function of $r$.

Using Eqs. (10-7) and (10-8), Eq. (10-6) becomes the same as Eq. (10-5) provided we are justified in neglecting the second and third terms on the right of Eq. (10-7), i.e., those containing $\nabla_\mu\Psi_R(r)$ and $\nabla_\mu^2\Psi_R(r)$. Therefore, we could expect the Born-Oppenheimer approximation to be valid if the electronic wavefunction is a slowly varying function of the nuclear coordinates. There is clear spectroscopic evidence to show that the approximation is indeed valid for the ground states of molecules, but that it is not so good for excited states of large polyatomic molecules. The approximation also breaks down in the case of degeneracy or near degeneracy of electronic wavefunctions.

The Born-Oppenheimer approximation gives electronic wavefunctions and energies for the various fixed positions of the nuclei. In the remainder of this book we shall assume that these are the wavefunctions and energies we are considering. We should also note that it is this approximation that allows us to draw potential energy curves of the type shown in Fig. 10-1 where we have plotted electronic energies for fixed nuclear positions.

## 10-3 THE MOLECULAR ORBITAL

Though you are probably quite familiar with the concept of the molecular orbital, we shall spend a moment clarifying this idea, because it is very important for an understanding of the contents of this chapter.

The orbital, atomic or molecular, is a one-electron function; it is an eigenfunction of a one-electron operator. For atoms, we defined the atomic orbital in terms of the eigenfunctions of the one-electron atom, i.e., the hydrogen-like atom. The physical significance of the atomic orbital is that it gives the probability of finding the electron that is assigned to

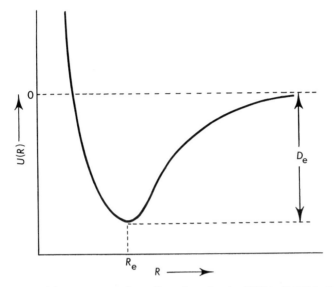

**Fig. 10-1**  Potential energy curve for a diatomic molecule. $U(R)(= E_e(R))$ is the potential energy for the nuclei. $R_e$ is the equilibrium internuclear distance; $D_e$ is the observed dissociation energy for the molecule plus the vibrational zero-point energy.

it. Though the orbital was defined for the one-electron atom, we saw that it was possible to extend the use of this concept by constructing many-electron functions from the one-electron functions.

These same ideas can be applied to molecules. Here we can start by defining molecular orbitals for a one-electron molecule. We can then extend the use by constructing molecular orbitals for many-electron molecules.

## 10-4  LINEAR COMBINATION OF ATOMIC ORBITALS

How do we construct molecular orbitals? The most generally used method is that called the *linear combination of atomic orbitals (LCAO) method*; this we shall now describe.

We recall (cf. Section 7-2) that one of the important methods for approximating solutions to Schroedinger equations is to use the variation theorem. We further recall that to do this in a systematic way it is necessary to choose a trial function with one or more parameters that can be varied to give the function corresponding to the lowest energy. In molecular problems we can obtain a trial function by taking a linear combination of the atomic orbitals of the atoms that form the molecule. If an electron in a molecule is near one of the nuclei, we can expect that the effect of this nucleus on it will be much larger than the effect of any of the other nuclei; consequently, it is reasonable for us to assume that the molecular orbital in which this electron finds itself is similar to the atomic orbital associated with the nucleus in question if it were not part of a molecule. In the LCAO method we assume that this approximation can be used as a starting point even when the electrons are not very near particular nuclei. The variation theorem then allows us to determine the best molecular orbitals that we can construct from a linear combination of the atomic orbitals.

If $\phi_i$ are the atomic orbitals, we can write a linear combination of these orbitals as follows:

$$\psi = a_1\phi_1 + a_2\phi_2 + \cdots + a_n\phi_n = \sum_i a_i\phi_i \qquad (10\text{-}9)$$

We now find the best values for the coefficients $a_i$ by applying the variation theorem as expressed in Eq. (7-1). In other words, we ask what values of the coefficients give the lowest value for the energy. Applying the theorem to the function in Eq. (10-9), we get

$$\frac{\int \psi^* \hat{H} \psi d\tau}{\int \psi^* \psi d\tau} \geqslant E_0 \qquad (10\text{-}10)$$

$\hat{H}$ is the correct Hamiltonian; $E_0$ is the lowest eigenvalue of this Hamiltonian.

Substitute Eq. (10-9) into Eq. (10-10); we get

$$\frac{\int \sum_i (a_i\phi_i) \hat{H} \sum_j (a_j\phi_j) d\tau}{\int \sum_i a_i\phi_i \sum_j a_j\phi_j d\tau} \geqslant E_0 \qquad (10\text{-}11)$$

where we have chosen the functions to be entirely real, which we can usually do. We can write Eq. (10-11) as

$$\frac{\sum_{ij} a_i a_j H_{ij}}{\sum_{ij} a_i a_j S_{ij}} \geqslant E_0 \qquad (10\text{-}12)$$

Here we are using the abbreviations

$$H_{ij} = \int \phi_i^* \hat{H} \phi_j d\tau \qquad (10\text{-}13)$$

$$S_{ij} = \int \phi_i^* \phi_j d\tau \qquad (10\text{-}14)$$

We call integrals like $S_{ij}$ *overlap integrals*; they play an important role in bonding theory. Integrals like $H_{ij}$ are often called *matrix elements* of the operator (in this case the Hamiltonian) because we can imagine a matrix of the integrals in which the $i$'s and $j$'s take all possible values.

We now have to minimize the left side of Eq. (10-12) by finding the best values for the $a$'s. If we define the left side of Eq. (10-12) as $E$, the energy associated with $\psi$, we get

$$E = \frac{\sum\limits_{ij} a_i a_j H_{ij}}{\sum\limits_{ij} a_i a_j S_{ij}} \tag{10-15}$$

or

$$\sum\limits_{ij} a_i a_j (H_{ij} - E S_{ij}) = 0 \tag{10-16}$$

Differentiate this with respect to one of the coefficients, say $a_k$, keeping the others constant; we get

$$\sum_i a_i H_{ki} + \sum_i a_i H_{ik} - E\left(\sum_i a_i S_{ki} + \sum_i a_i S_{ik}\right) - \frac{\partial E}{\partial a_k} \sum_{ij} a_i a_j S_{ij} = 0 \tag{10-17}$$

For the energy to be a minimum $\partial E / \partial a_k = 0$. Also, since we have assumed real functions, $S_{ki} = S_{ik}$, and the Hermitian condition makes $H_{ki} = H_{ik}$. Therefore,

$$\sum_i a_i (H_{ki} - E S_{ki}) = 0 \qquad (k = 1, 2, \ldots, n) \tag{10-18}$$

There will be $n$ equations of this type, one for each coefficient that can be varied. Therefore, we have obtained a set of $n$ linear equations in $n$ unknowns.

Before continuing, we shall digress to explain how sets of linear equations, such as we have here, can be solved.

## Solution of Sets of Linear Equations

Consider the following set of $n$ linear equations in $n$ unknowns:

$$C_{11}x_1 + C_{12}x_2 + \cdots + C_{1n}x_n = y_1$$
$$C_{21}x_1 + C_{22}x_2 + \cdots + C_{2n}x_n = y_2$$
$$\cdots\cdots\cdots\cdots\cdots\cdots\cdots \tag{10-19}$$
$$C_{n1}x_1 + C_{n2}x_2 + \cdots + C_{nn}x_n = y_n$$

We can solve these equations by elimination. But we can also solve them using determinants. We shall not give a complete analysis of the solutions of sets of linear equations, but shall consider that part of the analysis that is relevant to the problem at hand. In what follows we shall make considerable reference to the part of Section 8-4 where we examined matrices and determinants; all the definitions and terminology we need are either there or here.

We can write Eqs. (10-19) in matrix form:

$$Cx = y \tag{10-20}$$

$C$ is the matrix of the coefficients; $x$ and $y$ are column matrices. We first consider the case where $y = 0$. Multiply Eq. (10-20) from the left by the inverse of $C$; we get

$$x = C^{-1}y = \frac{\hat{C}y}{|C|} \tag{10-21}$$

$|C|$ is the determinant of the coefficients. By the rule for matrix multiplication and the definition of the adjugate matrix, we can write

$$x_i = \frac{1}{|C|}(y_1 C^{1i} + y_2 C^{2i} + \cdots + y_n C^{ni}) \tag{10-22}$$

where $C^{ji}$ is the cofactor of the element $C_{ji}$ in $|C|$. This method for solving a set of linear equations is called *Cramer's rule*. For the method to work $|C|$ must obviously be non-zero.

Now consider the situation if all the $y$'s in Eqs. (10-19) are zero, i.e., we have a set of *homogeneous* linear equations — Eq. (10-18) is of this type. We further assume that $|C|$ is non-zero. From Eq. (10-22) we see that in this case all the $x$'s are zero. We call this a *trivial* solution. Obviously, it is neither very interesting, nor very useful.

The question we now ask is: What is the necessary condition for a set of homogeneous linear equations to have a non-trivial solution?

We first give two definitions: The *rank* of a matrix is the order (number of rows or columns) of the largest non-zero determinant contained in it. The *augmented matrix* for a set of linear equations is the matrix of the coefficients to which we add a column containing the constants in the equations; for example, consider the following very simple case:

$$3x + 2y + 2z = 13$$
$$2x - y - z = -3$$
$$4x + 5y - 3z = 5$$

The matrix of the coefficients $C$ and the augmented matrix $A$ are

$$C = \begin{bmatrix} 3 & 2 & 2 \\ 2 & -1 & -1 \\ 4 & 5 & -3 \end{bmatrix} \qquad A = \begin{bmatrix} 3 & 2 & 2 & 13 \\ 2 & -1 & -1 & -3 \\ 4 & 5 & -3 & 5 \end{bmatrix}$$

Here rank $C$ = rank $A$ = 3, and you can easily check that we can solve for the three unknowns. This is generally true: If the rank of the matrix of the coefficients is equal to the rank of the augmented matrix and if the rank is $n$, the number of unknowns, then we can solve for these $n$ unknowns; there is a single solution which we can obtain by using Cramer's rule.

When we have a set of $n$ homogeneous equations in $n$ unknowns, rank $C$ = rank $A$; this is because $A$ differs from $C$ only by having an additional column of zeros. If the rank equals the number of unknowns, then $|C|$ is non-zero, and we have already seen that in this case the solution is the trivial one. For a non-trivial solution we must therefore have rank $C < n$ (it obviously cannot be greater than $n$). This means that $|C| = 0$ because the largest non-zero determinant is smaller than the determinant of the coefficients. In words, we can say that a set of $n$ homogeneous equations in $n$ unknowns has a non-trivial solution only if the determinant of the coefficients is zero. This is a very important result—we shall now use it to continue our discussion of the LCAO method for determining molecular orbitals.

From the above discussion we can conclude that Eq. (10-18) has a non-trivial solution only if the determinant of the factors that multiply the unknowns $a_i$ is zero, i.e.,

$$\begin{vmatrix} H_{11} - ES_{11} & H_{12} - ES_{12} & \cdots & H_{1n} - ES_{1n} \\ H_{21} - ES_{21} & H_{22} - ES_{22} & \cdots & H_{2n} - ES_{2n} \\ \cdots\cdots\cdots\cdots\cdots\cdots\cdots\cdots\cdots\cdots\cdots\cdots\cdots \\ H_{n1} - ES_{n1} & H_{n2} - ES_{n2} & \cdots & H_{nn} - ES_{nn} \end{vmatrix} = 0 \qquad (10\text{-}23)$$

In abbreviated form we write this

$$|H_{ki} - ES_{ki}| = 0 \qquad (10\text{-}24)$$

We call the determinant in Eq. (10-23) or Eq. (10-24) a *secular determinant*. When we expand it we get an $n$th degree polynomial in $E$. The lowest of the $n$ values for $E$ is the best value for the ground state energy obtainable from a function of type Eq. (10-9). To determine the function corresponding to this energy we substitute the value for $E$ into Eq. (10-18) and solve for the coefficients to be put into Eq. (10-9)—this procedure gives us the ratios of the coefficients to be put into Eq. (10-9), and normalization of the resulting function completes the determination of its form. The other $(n-1)$ values we calculate for $E$ are approximations to the energies of functions corresponding to excited states; we determine the forms of these functions in the same way as for the ground state function. In the next section we see how we can apply this theory to a one-electron molecule in order to find molecular orbitals.

## 10-5   THE HYDROGEN MOLECULE ION

The simplest molecule is the singly-ionized hydrogen molecule shown schematically in Fig. 10-2. This molecule is, admittedly, not particularly interesting as a chemical species. However, as a starting point for discussion of bonding in many-electron molecules, it is very important— it plays a somewhat similar role in molecular bonding theory to the role played by the hydrogen atom in the understanding of atomic theory.

If we set up the non-relativistic Schroedinger equation for the molecule and use the Born-Oppenheimer approximation of fixed nuclei, it is possible to solve the equation exactly. To do this it is necessary to use a special coordinate system. The Schroedinger equation for the hydrogen

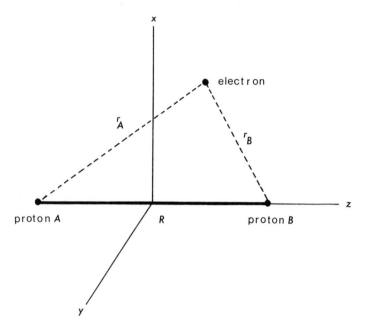

**Fig. 10-2**   Distances used in the $H_2^+$ problem.

molecule ion is not separable in cartesian coordinates, but it is separable in what are known as *confocal elliptical coordinates* — this latter co-ordinate system is defined by

$$\xi = \frac{r_A + r_B}{R} \qquad \eta = \frac{r_A - r_B}{R} \tag{10-25}$$

and $\phi$, the angle of rotation about the internuclear axis. We shall not, however, pursue the exact solution since it is rather long, and further-more it does not give a general method for solution of molecular bonding problems†. In fact, we have a similar situation to that which we met when dealing with atoms: the single electron atomic problem can be solved exactly, but many-electron atoms have to be dealt with by approximation techniques. Similarly, for many-electron molecules we have to use approximation techniques. We shall therefore treat the simplest molecular bonding problem by the approximation method that is commonly used

†For more details of the exact solution of the $H_2^+$ problem see for example, Slater, J. C., *Quantum Theory of Molecules and Solids*, Vol. 1, McGraw-Hill, New York, 1963, p. 247.

for more complex molecules, the LCAO method that we described in the last section. But before we do this we shall make some preliminary comments about the system we are examining.

Within the limits of the Born-Oppenheimer approximation, the system is an electron in motion about two fixed protons (Fig. 10-2). The Hamiltonian we shall work with is therefore (in a.u.)

$$\hat{H} = -\frac{1}{2}\nabla^2 - \frac{1}{r_A} - \frac{1}{r_B} \qquad (10\text{-}26)$$

We have not included in $\hat{H}$ the term associated with the internuclear repulsion. This latter term is equal to $1/R$ — for fixed nuclear positions it is obviously constant, and can be taken into account later.

The Schroedinger equation we have to solve can be written:

$$\hat{H}\psi_n = E_n\psi_n \qquad (10\text{-}27)$$

In this equation $\hat{H}$, $\psi_n$, and $E_n$ all depend on the internuclear distance $R$. When $R = 0$, the system becomes a singly-ionized helium atom, for then the two protons have coalesced into a single nucleus of charge 2. We have determined the energies and wavefunctions for such a system before (cf. Eqs. (6-119) and (6-120)). The energies are given by $E_n = -Z^2/2n^2$ hartrees. For the ground state $n = 1$, and $E_1 = -2$ hartrees. On the other hand, when $R \rightarrow \infty$, the system becomes a hydrogen atom and a bare proton, $\psi_n$ is then just the hydrogen atom wavefunction, and $E_n = -1/2n^2$ hartrees (cf. Eq. (6-120)). For the ground state the energy is now given by $E_1 = -0.5$ hartrees. We can conclude therefore that the limiting values of $E_1$ are $-2$ hartrees and $-0.5$ hartrees. What we are really interested in, however, is the solution to Eq. (10-27) for all values of $R$, especially those corresponding to any bond that may be formed in the system. In this regard, we must consider the total molecular energy for the various fixed nuclear positions. If we designate this $E(R)$, we can obtain it by adding the internuclear repulsion to the energy $E_1$, i.e.,

$$E(R) = E_1 + \frac{1}{R} \qquad (10\text{-}28)$$

In Fig. 10-3 we have illustrated how the quantities in Eq. (10-28) vary with $R$. From this diagram we see that there is a minimum in the curve for the total molecular energy — this indicates that a bond does indeed form. The minimum occurs at an equilibrium internuclear distance of 2 bohrs, and leads to a dissociation energy of 0.1 hartrees [11].

We now proceed to apply the LCAO method (cf. Section 10-4) to the

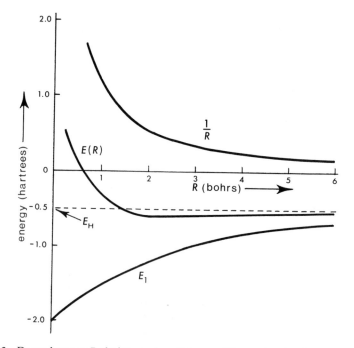

**Fig. 10-3** Dependence on $R$, the internuclear distance, of the total molecular energy for $H_2^+$, $E(R)$, $E_1$ and the internuclear repulsion $1/R$. $E_H$ is the energy of an electron in a $1s$ atomic orbital of hydrogen.

hydrogen molecule ion problem. By so doing, we can determine approximations for the $E_n$ and $\psi_n$ that occur in Eq. (10-27).

If the electron is near one of the protons, we would expect that the orbital associated with it would be very similar to a $1s$ hydrogen atom orbital. In our linear combination of atomic orbitals to form molecular orbitals we shall therefore use two $1s$ atomic orbitals; we label these $1s_A$ and $1s_B$. To determine the energies $E_n$ we have to write the secular determinant. In this case the equation we have to solve is (cf. Eq. (10-23))

$$\begin{vmatrix} H_{AA}-E_n & H_{AB}-E_nS \\ H_{BA}-E_nS & H_{BB}-E_n \end{vmatrix} = 0 \tag{10-29}$$

where

$$H_{AA} = \int 1s_A \hat{H} 1s_A d\tau \tag{10-30}$$

$$H_{BB} = \int 1s_B \hat{H} 1s_B d\tau = H_{AA} \tag{10-31}$$

$$H_{AB} = \int 1s_A \hat{H} 1s_B d\tau \tag{10-32}$$

$$H_{BA} = \int 1s_B \hat{H} 1s_A d\tau = H_{AB} \tag{10-33}$$

$$S = S_{AB} \quad \text{or} \quad S_{BA} = \int 1s_A 1s_B d\tau = \int 1s_B 1s_A d\tau \tag{10-34}$$

You should note that we have not written $S_{AA}$ or $S_{BB}$ in Eq. (10-29) since these are one due to our using the normalized $1s$ atomic orbitals. (However, $S_{AB}$ and $S_{BA}$ are not zero because the two atomic orbitals, one from each atom, are not orthogonal to each other.)

We shall not work through the details of solving Eq. (10-29) since it would be well for you to do this as an exercise (Exercise 10-1); it is in any case just a simple algebraic problem. The result is two possible values for the energy:

$$E_g = \frac{H_{AA} + H_{AB}}{1 + S} \tag{10-35}$$

$$E_u = \frac{H_{AA} - H_{AB}}{1 - S} \tag{10-36}$$

We shall explain the $g$ and $u$ subscripts in a moment. In the meantime we shall consider the forms of the molecular orbitals corresponding to $E_g$ and $E_u$. These are determined by substituting the energies into Eq. (10-18). If we did this, we would determine that for $E_g$, $a_1 = a_2$ (cf. Eq. (10-9)), and for $E_u$, $a_1 = -a_2$. Therefore, after normalization, the molecular orbitals are given by

$$\psi_g = \frac{1s_A + 1s_B}{(2 + 2S)^{1/2}} \tag{10-37}$$

$$\psi_u = \frac{1s_A - 1s_B}{(2 - 2S)^{1/2}} \tag{10-38}$$

The factors $(2 + 2S)^{-1/2}$ and $(2 - 2S)^{-1/2}$ are normalization constants. In Eqs. (10-35) to (10-38) we have used the subscripts $g$ and $u$ to indicate the symmetry type of the orbitals with respect to inversion through the mid-point between the nuclei. In the coordinate system of Fig. 10-2, this operation can be described as follows: $x \rightarrow -x$, $y \rightarrow -y$ and $z \rightarrow -z$. A $g$ subscript indicates that the orbital function does not change when this inversion operation is performed, i.e., the orbital is symmetric; a $u$ subscript indicates that the orbital function changes sign when the inversion operation is performed; i.e., the orbital is antisymmetric.

In Fig. 10-4 we have shown in a very schematic way what the orbitals

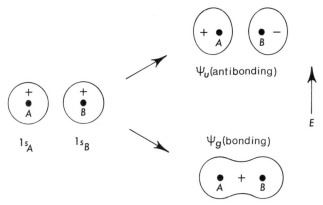

**Fig. 10-4**   Diagrammatic representation of the $\psi_g$ and $\psi_u$ molecular orbitals for $H_2^+$.

$\psi_g$ and $\psi_u$ look like. Another way of examining these orbitals is to consider a plot of the electron probability density $|\psi|^2$ along the line joining the nuclei—plots of this type for the two orbitals are shown in Fig. 10-5. From Fig. 10-4 and Fig. 10-5 we see that the $\psi_g$ orbital corresponds to a large electron probability density between the nuclei, whereas the $\psi_u$ orbital corresponds to a relatively much lower probability density between the nuclei. In fact, $\psi_g$ is a bonding orbital because it allows the electron a large probability of being in the region of space where it comes under the influence of both nuclei; $\psi_u$ is an antibonding orbital due to the fact that the electron has a relatively low probability of being in this region of space. Evaluation of the energies $E_g$ and $E_u$, as given by Eqs. (10-35) and (10-36), would confirm these qualitative findings. If we did evaluate the integrals in Eqs. (10-35) and (10-36), we would discover that both $H_{AA}$ and $H_{AB}$ are negative; therefore $E_g$ is a lower energy than $E_u$, which means that $\psi_g$ is the ground state of the system.

We shall now examine the energies $E_g$ and $E_u$ in a little more detail. To do this we need to consider the integrals $S$, $H_{AA}$, and $H_{AB}$.

$S$ is the overlap integral, given by Eq. (10-34). Using the normalized $1s$ atomic orbitals (cf. Eq. (6-119)), Eq. (10-34) becomes

$$S = \frac{1}{\pi} \int e^{-(r_A + r_B)} d\tau \qquad (10\text{-}39)$$

We can evaluate this integral using the elliptical coordinates defined in Eq. (10-25); if we did this, we would determine $S$ to be given by (Exercise 10-3)

$$S = e^{-R}\left(1 + R + \frac{R^2}{3}\right) \qquad (10\text{-}40)$$

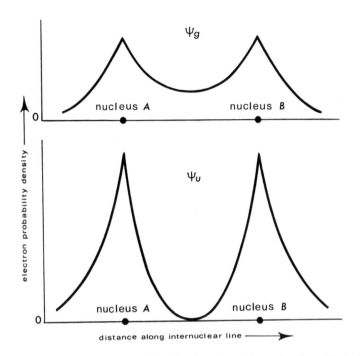

**Fig. 10-5** Electron probability densities for the $\psi_g$ and $\psi_u$ molecular orbitals for $H_2^+$ plotted along the line joining the nuclei.

We therefore see that when $R$ becomes large, $S$ approaches zero — this is when there is no significant overlap between the atomic orbitals. As $R$ approaches zero, $S$ approaches unity as $r_A$ and $r_B$ become identical.

To examine the integral $H_{AA}$ we need to refer to the Hamiltonian in Eq. (10-26). This integral is then given by

$$H_{AA} = \int 1s_A \left( -\frac{1}{2}\nabla^2 - \frac{1}{r_A} - \frac{1}{r_B} \right) 1s_A d\tau$$

i.e.,

$$H_{AA} = \int 1s_A \left( -\frac{1}{2}\nabla^2 - \frac{1}{r_A} \right) 1s_A d\tau + \int 1s_A \left( -\frac{1}{r_B} \right) 1s_A d\tau$$

$$= E_H + \int 1s_A \left( -\frac{1}{r_B} \right) 1s_A d\tau \tag{10-41}$$

Here $E_H$ is the energy of an electron in a $1s$ orbital of the hydrogen atom, i.e., $-0.5$ hartrees. The second term on the right of Eq. (10-41) can be given an electrostatic interpretation. It represents the electrostatic potential at nucleus $B$ that would occur if there was a spherically symmetric charge distribution, with charge density $(1s_A)^2$, about nucleus $A$. Thus $H_{AA}$ can be considered to be the energy of an electron in a hydrogen atom in its ground state plus the energy due to the attraction of a proton $B$ for this electron. If a hydrogen atom and a proton are brought together, $H_{AA}$ tends to stabilize the system since the second term in Eq. (10-41) becomes more negative as $R$ decreases. However, this stabilization is cancelled by the internuclear repulsion. To account for the bonding in the hydrogen molecule ion we must therefore look to the integral $H_{AB}$.

Before we consider $H_{AB}$ we should mention that if we were to evaluate $H_{AA}$, we would determine it to be given by

$$H_{AA} = -\frac{1}{2} - \frac{1}{R} + e^{-2R}\left(1 + \frac{1}{R}\right) \qquad (10\text{-}42)$$

This expression indicates the dependence of $H_{AA}$ on $R$. At large values of $R$, the integral tends to $-0.5$ hartrees, the energy of a hydrogen atom in its ground state – this is what we would expect in the light of the comments we have just made.

The integral $H_{AB}$ takes account of the fact that the electron is not restricted to the orbitals $1s_A$ or $1s_B$ associated with nuclei $A$ and $B$, but can undergo exchange between the two orbitals. It is given by

$$H_{AB} = \int 1s_A \left(-\frac{1}{2}\nabla^2 - \frac{1}{r_A} - \frac{1}{r_B}\right) 1s_B d\tau$$

$$= E_H S + \int 1s_A \left(-\frac{1}{r_A}\right) 1s_B d\tau \qquad (10\text{-}43)$$

Evaluation of $H_{AB}$ gives

$$H_{AB} = -\frac{S}{2} - e^{-R}(1 + R) \qquad (10\text{-}44)$$

From Eq. (10-44) we see that for large values of $R$, $H_{AB}$ is very small. However, at values of $R$ corresponding to the bond formation $H_{AB}$ is appreciable; it is negative, and therefore accounts for the binding energy of the molecule.

The total molecular energy for the ground state of the molecule is given by substituting Eqs. (10-40), (10-42) and (10-44) into Eq. (10-35) and adding the nuclear repulsion energy (cf. Eq. (10-28)). The discussions of the past few paragraphs have shown that this total energy is strongly dependent on $R$, the internuclear distance. In Fig. 10-6 we have plotted

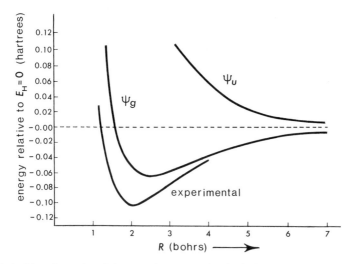

**Fig. 10-6**   Plot of energies of the $\psi_g$ and $\psi_u$ states of $H_2^+$ relative to $E_H = 0$. The internuclear distance and the energies are in atomic units.

the energy for the $\psi_g$ orbital against $R$, using the energy of the hydrogen atom $E_H$ as a reference point. This diagram shows that $\psi_g$ is a bonding orbital. The calculated dissociation energy is 0.065 hartrees which occurs at an equilibrium internuclear distance of 2.5 bohrs. We must compare these values to the correct ones of 0.10 hartrees and 2 bohrs. Obviously, they are not all that could be desired. However, the theory does account for a major part of the binding energy of the molecule. Furthermore, it can be applied to more complex molecules—this latter point is very important.

There is one further point to make before we leave the hydrogen molecule ion problem. In Fig. 10-6 we have also shown how the energy of the $\psi_u$ orbital is dependent on $R$. We see that this orbital is antibonding, and if the electron is assigned to it the system is in a repulsive state.

## 10-6   THE HYDROGEN MOLECULE

Examination of the bonding in the hydrogen molecule is very important for an understanding of chemical bonding in general; this is because in the hydrogen molecule we have the simplest two-electron bond. If we can understand the nature of this bond in hydrogen, we can extend what we learn to more complex molecules. Naturally, therefore, much theoretical research has been done on the hydrogen molecule†. If we wanted to examine this research chronologically, we would first have to look at the valence-bond treatment of the molecule. We shall not do this, but instead consider the molecular-orbital treatment — we examine the valence-bond treatment in the next chapter.

The hydrogen molecule, with the notation we shall use, is shown in Fig. 10-7.

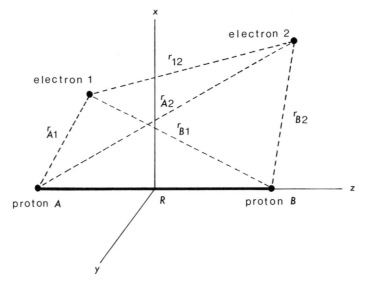

**Fig. 10-7**    Distances used in the hydrogen molecule problem.

The non-relativistic Hamiltonian for the system is

$$\hat{H} = \left(-\frac{1}{2}\nabla_1{}^2 - \frac{1}{r_{A1}} - \frac{1}{r_{B1}}\right) + \left(-\frac{1}{2}\nabla_2{}^2 - \frac{1}{r_{A2}} - \frac{1}{r_{B2}}\right) + \frac{1}{r_{12}} \qquad (10\text{-}45)$$

We have omitted the nuclear repulsion term $1/R$ from this Hamiltonian — as we did in the $H_2{}^+$ problem; this is a constant for fixed nuclei and can be accounted for later. The reason why we have expressed the Hamiltonian like this is to show that it can be written as the sum of two $H_2{}^+$ Hamiltonians and the electron-electron interaction term (cf. Eq. (10-26)),

---

†For references to and summary of this work see, for example, reference C2 in the Bibliography.

i.e.,

$$\hat{H} = \hat{H}'(1) + \hat{H}'(2) + \frac{1}{r_{12}} \qquad (10\text{-}46)$$

Writing $\hat{H}$ like this suggests a method for approximating the solution.

To approximate the solution of the Schroedinger equation for the hydrogen molecule we omit the last term, i.e., the electron-electron interaction term. Since this term is important, we would expect that its omission would cause a significant error in our calculations. Nevertheless, we proceed without it, and consider it later. Let us denote the Hamiltonian we shall use by $\hat{H}^0$. The equation we wish to solve is then

$$\hat{H}^0\Psi = E\Psi$$

i.e.,

$$[\hat{H}'(1) + \hat{H}'(2)]\Psi = E\Psi \qquad (10\text{-}47)$$

This equation is now susceptible to solution by the separation of variables technique. If this is done (Exercise 10-4), we get the solution

$$\Psi = \psi_g(1)\psi_g(2) \qquad (10\text{-}48)$$

where $\psi_g$ is the familiar $H_2^+$ molecular orbital given by Eq. (10-37), i.e.,

$$\psi_g = \frac{1s_A + 1s_B}{(2+2S)^{1/2}}$$

To write the complete function we must include the electron spin. Because we are now dealing with two electrons we must ensure that the Pauli principle is obeyed. The complete, normalized function is therefore

$$\Psi_{MO} = \psi_g(1)\psi_g(2)\frac{1}{\sqrt{2}}[\alpha(1)\beta(2) - \beta(1)\alpha(2)] \qquad (10\text{-}49)$$

We should note that the spatial part of this function is symmetric, whereas the spin part is antisymmetric—this gives a complete function that is antisymmetric, in accord with the requirements of the Pauli principle.

We could now calculate the energy of the molecule in a similar way to that used for the helium atom (cf. Section 9-4). We therefore use our approximate wavefunction $\Psi_{MO}$ and the correct Hamiltonian given in Eq. (10-45). Remembering that the total molecular energy $E(R)$ includes the internuclear repulsion, we get†

$$E(R) = \int \Psi_{MO}\hat{H}\Psi_{MO}d\tau + \frac{1}{R} \qquad (10\text{-}50)$$

†For a more detailed treatment of the simple MO treatment of the hydrogen molecule see, for example, reference C2 in the Bibliography, p. 479ff.

The approximation we have used here leads to a dissociation energy for the hydrogen molecule of 0.0974 hartrees at an internuclear distance of 1.57 bohrs [12]. We should compare these values with the experimental ones of 0.174 hartrees and 1.40 bohrs.

The results obtained from this treatment show that the function Eq. (10-49) is not a very good one. However, it does account for a large portion of the binding energy of the molecule: considering the simplicity of the treatment, it is not an unsatisfactory result. Later in this chapter we shall briefly discuss how the MO treatment of the hydrogen molecule can be improved. In the meantime, we consider how the simple application of MO theory we have examined in Sections 10-5 and 10-6 can be extended to the qualitative understanding of the bonding in other diatomic molecules.

## 10-7 MO THEORY FOR MORE COMPLEX DIATOMIC MOLECULES

Having dealt with the simplest molecular systems, we now naturally turn to the next most complex — diatomic molecules with more than two electrons. Here we consider in a qualitative manner how we can extend the ideas of simple MO theory to these systems — we consider how we apply the LCAO theory and how we label the molecular orbitals and states that result.

When we constructed a linear combination of the hydrogen atom $1s$ orbitals we obtained two molecular orbitals — one was symmetric for the inversion operation; the other was antisymmetric. We can expect that two $2s$ orbitals will combine similarly; again we shall get symmetric and antisymmetric molecular orbitals.

In Figs. 10-8a and 10-8b we show schematically how $2p$ orbitals combine. Here we have two sets of atomic orbitals, each consisting of three orbitals, so we expect six molecular orbitals to be formed. The atomic orbitals that project along the internuclear axis will give a different type of molecular orbital to the other two pairs. We can determine whether these orbitals are bonding or antibonding by considering the electron probability density between the nuclei: a large probability density corresponds to a bonding situation; a low probability density to an antibonding situation. We shall not pursue these matters further since you are probably quite familiar with them from your introductory chemistry courses.

We now examine the labeling of molecular orbitals for diatomic

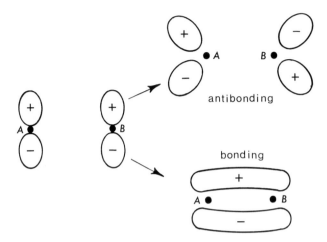

**Fig. 10-8a** Diagrammatic representation of molecular orbitals formed from $2p$ atomic orbitals perpendicular to the molecular axis.

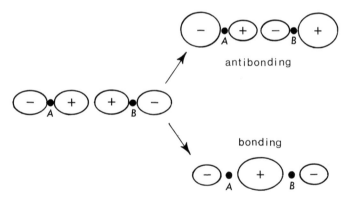

**Fig. 10-8b** Diagrammatic representation of molecular orbitals formed from $2p$ atomic orbitals that lie along the molecular axis.

molecules. As with atomic orbitals, we use the angular momentum to label them and define an analogous quantum number to $m$—we give it the symbol $\lambda$. $\lambda$ refers to the component of the orbital angular momentum in the direction of the internuclear axis. Again, analogously to using the letters $s, p, d, f$, etc. for atomic orbitals, for diatomic molecules we use the Greek letters $\sigma, \pi, \delta, \phi$, etc. to label molecular orbitals with $|\lambda|$ values $0, 1, 2, 3$, etc.

In the case of homonuclear diatomic molecules we have already seen that the orbitals are either symmetric or antisymmetric to inversion through the center of symmetry. We can therefore further characterize our orbitals for these molecules with $g$ or $u$ symbols — we write, for example, $\sigma_u$ or $\pi_g$. Heteronuclear diatomic molecules do not have this type of symmetry, so the $g$ and $u$ symbols are not used for them.

A further label is needed to distinguish molecular orbitals with the same value for $|\lambda|$ and the same inversion symmetry. Several methods are available for doing this. For example, we can attach a label that indicates the atomic orbitals that the molecular orbital would dissociate into if we pulled the nuclei apart. Thus, we can label the orbitals $\sigma_g 1s$, $\sigma_u^* 1s$, $\sigma_g 2s$, $\sigma_u^* 2s$, $\pi_g^* 2p$, $\pi_u 2p$, and so on. (The asterisks denote antibonding orbitals; orbitals without asterisks are bonding.) In this notation, a $\sigma_g 2s$ orbital, for example, is one that can be considered to be built up from two $2s$ atomic orbitals. Another, and simpler, method for distinguishing orbitals with the same value for $|\lambda|$ and the same inversion symmetry is to number them in order of increasing energy. Thus, we get orbitals labeled $1\sigma_g$, $2\sigma_g$, $3\sigma_g$, ..., $1\sigma_u$, $2\sigma_u$, $3\sigma_u$, ..., $1\pi_g$, $2\pi_g$, ..., etc. In this book we shall use this latter notation.

To determine the ground state molecular electronic configurations, we place the electrons into the molecular orbitals that lead to the lowest energy for the molecule under consideration; when we do this we must, of course, ensure that the Pauli principle is not violated. This method for determining molecular electronic configurations is similar to the aufbau process used to determine atomic electronic configurations. We therefore need to know the relative order of energies of the available molecular orbitals. In general this is determined by the particular atomic orbitals from which the various molecular orbitals are formed and by the extent of overlap of these atomic orbitals. A typical order is shown in Fig. 10-9. We should note that the $\pi$-orbitals are doubly degenerate. We should also note that the orbitals $1\sigma_g$, $2\sigma_g$, $1\pi_u$, and $3\sigma_g$ are bonding orbitals, whereas the orbitals $1\sigma_u$, $2\sigma_u$, $1\pi_g$ and $3\sigma_u$ are antibonding orbitals.

The configurations of the ground states of the possible homonuclear diatomic molecules of the elements up to fluorine are given in Table 10-1. In this table we use the symbol $K$ to indicate that the inner electrons are essentially still in $1s$ atomic orbitals; i.e., $K$ means two electrons in a $1s$ orbital — we expect this to occur whenever the overlap between the $1s$ atomic orbitals is slight. We should note that except for $He_2$ and $Be_2$ there is an excess of bonding over antibonding electrons — this accounts for the stability of these molecules. In $He_2$ and $Be_2$ the number of bonding electrons equals the number of antibonding electrons; in general, when

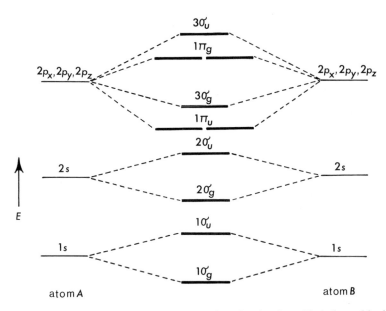

**Fig. 10-9** Schematic representation of the energies of molecular orbitals formed by linear combination of atomic orbitals. The order of energies shown is typical for homonuclear molecules formed from elements in the first row of the periodic table. The relative spacing between the levels is arbitrary.

this is the case instability results—$He_2$ and $Be_2$ are not stable in their ground states.

Now that we have considered the individual orbitals into which the electrons of diatomic molecules can go, we should look briefly at the total electronic picture. When we discussed this subject with regard to atoms we used term symbols to indicate the total electronic states. We can do the same for molecules. Here again, though, it is the total orbital angular momentum component along the internuclear axis that is important. This is labeled by $\Lambda$, and depends on the contributions from the individual electrons. What we do is sum the individual $\lambda$ values—this gives the molecular term symbol, as follows:

$$|\Lambda|: \qquad 0 \quad 1 \quad 2 \quad 3 \quad \cdots$$
$$\text{Term symbol: } \Sigma \quad \Pi \quad \Delta \quad \Phi \quad \cdots$$

(Note that the general scheme is to use Roman letters for atoms and Greek letters for molecules—small letters refer to individual electron properties; capital letters refer to total electron properties.)

**Table 10-1**   Electronic configurations and term symbols for some diatomic molecules.

| Molecule | Configuration† | Ground state term symbol |
|---|---|---|
| $H_2$ | $(1\sigma_g)^2$ | $^1\Sigma_g^+$ |
| $He_2$ (ground state not stable) | $(1\sigma_g)^2 (1\sigma_u)^2$ | $^1\Sigma_g^+$ |
| $Li_2$ | $KK(2\sigma_g)^2$ | $^1\Sigma_g^+$ |
| $Be_2$ (ground state not stable) | $KK(2\sigma_g)^2 (2\sigma_u)^2$ | $^1\Sigma_g^+$ |
| $B_2$ | $KK(2\sigma_g)^2 (2\sigma_u)^2 (1\pi_u)^2$ | $^3\Sigma_g^-$ |
| $C_2$ | $KK(2\sigma_g)^2 (2\sigma_u)^2 (1\pi_u)^4$ | $^1\Sigma_g^+$ |
| $N_2$ | $KK(2\sigma_g)^2 (2\sigma_u)^2 (1\pi_u)^4 (3\sigma_g)^2$ | $^1\Sigma_g^+$ |
| $O_2$ | $KK(2\sigma_g)^2 (2\sigma_u)^2 (1\pi_u)^4 (3\sigma_g)^2 (1\pi_g)^2$ | $^3\Sigma_g^-$ |
| $F_2$ | $KK(2\sigma_g)^2 (2\sigma_u)^2 (1\pi_u)^4 (3\sigma_g)^2 (1\pi_g)^4$ | $^1\Sigma_g^+$ |

†$KK$ represents 4 electrons in $1s$ atomic orbitals.

Spin angular momentum for atoms manifests itself in the multiplicity of the atomic state, written as a left superscript on the term symbol. Similarly for molecules, we write the total spin multiplicity of a molecule's state $(= 2S + 1)$ as a left superscript.

For diatomic molecules there are some symmetry operations that allow further classification of the total electronic states. One symmetry operation that can be performed on homonuclear diatomic molecules is inversion through the center of symmetry. We label states that are symmetric with respect to this operation with a $g$; states that are antisymmetric are labeled $u$. To determine whether a state is $g$ or $u$ we multiply the inversion symmetries of all the individual electrons: $g \times g = u \times u = g$; $g \times u = u$.

The other symmetry operation of interest to us here is reflection in a plane containing the internuclear axis. Though we can perform this symmetry operation for $\Pi$, $\Delta$, $\Phi$, etc. states, for reasons that we shall not go into here, it is only for $\Sigma$ states that it may be informative†. We label these latter states with a $+$ superscript if the reflection operation is symmetric, or with a $-$ superscript if it is antisymmetric.

†If you are interested in the details, you can consult reference F12 in the Bibliography, p. 129.

## 10-8  IMPROVEMENTS ON THE SIMPLE MO METHOD

We saw in Section 10-6 that the very simple MO treatment of the hydrogen molecule did not give very good results. In the last section we have seen that the simple treatment can be extended to more complex molecules. Considering the extent of the inadequacy of the simple treatment for the hydrogen molecule, we might expect that for both this molecule and for more complex molecules considerably more sophistication would be needed to obtain satisfactory results. In fact, there are several improvements that can be effected in the MO method. It is well beyond the scope of this book to deal with these improvements. However, we should very briefly indicate the direction these improvements take. The discussion will be very cursory; for more details you can consult one of the references in the Bibliography, e.g., reference C2.

Though the material in this section is generally applicable to molecular bonding problems, we shall use the hydrogen molecule as an example. First, we consider the wavefunction we obtained for this molecule using the simple MO treatment — this is given by Eq. (10-49). What is wrong with this function? We can get an indication of the trouble by expanding the spatial part; doing this, we get

$$\frac{1}{(2+2S)^{1/2}}[1s_A(1)+1s_B(1)]\frac{1}{(2+2S)^{1/2}}[1s_A(2)+1s_B(2)]$$
$$=\frac{1}{2+2S}[1s_A(1)1s_A(2)+1s_A(1)1s_B(2)$$
$$+1s_B(1)1s_A(2)+1s_B(1)1s_B(2)] \qquad (10\text{-}51)$$

The second and third terms in this expansion represent situations with one electron on each nucleus; i.e., situations where the electrons are shared by the nuclei. We can consider these to be covalent terms. The first and last terms correspond to both electrons being on the same nucleus; we can consider these to be ionic structures $H_A^-H_B^+$ and $H_A^+H_B^-$. Furthermore, the function of Eq. (10-51) gives equal weight to the ionic and covalent terms — we can therefore guess that this is the trouble with it, because intuitively we always regard the bonding in hydrogen to be almost entirely covalent.

The discussion of the last paragraph suggests that we might improve the function by omitting the ionic terms. We shall not discuss the details here because by so doing we get a function that is equivalent to that used in the valence-bond treatment of the hydrogen molecule, which we take up in the next chapter. Instead, we shall look at other methods of improving the function.

One of the problems with the function in Eq. (10-49) is that it is constructed from a very restricted basis set of atomic orbitals. In fact, to obtain it we used only the two $1s$ atomic orbitals of the constituent hydrogen atoms. We can improve the MO function by using more atomic orbitals (e.g., the $2s$ and $2p_z$ orbitals) in the linear combination of atomic orbitals. This has the effect of increasing the basis set and introducing more variation parameters.

To apply the variation theorem to the solution of the hydrogen molecule and other molecular bonding problems in a more systematic way, it is possible to apply a Hartree-Fock self-consistent field treatment similar to that used for the closed-shell atomic problem (cf. Section 9-11). When we do this the equations to be solved are analogous to Eq. (9-46); they have the form

$$\hat{H}_{\text{eff}}\psi_i(i) = E_i\psi_i(i) \tag{10-52}$$

Now the $\psi_i$ are molecular orbitals instead of the atomic orbitals in Eq. (9-46). (The molecular orbitals may be formed by the LCAO method.) $\hat{H}_{\text{eff}}$, as in the atomic problem, includes a potential energy term $\hat{U}_{\text{av}}$ that accounts for the average interelectronic interaction. The equations in Eq. (10-52) are solved by iterative methods similar to those described for the atomic problem. If the variation of the $\psi_i$ is complete (with the condition that the $\psi_i$ remain as an orthonormal set), and the iteration is carried out until self-consistency is reached, the molecular orbitals are the *Hartree-Fock orbitals*.

The Hartree-Fock wavefunctions are the best wavefunctions of the independent-particle type. However, again referring to the atomic problem (cf. Section 9-12) we know that these wavefunctions do not correlate the electrons' motions completely. We recall that the Hartree-Fock wavefunctions tend to keep electrons with like spin away from each other, i.e., to correlate their motions; however, this type of function does not adequately correlate the motions of electrons with unlike spin.

One way of dealing with the electron correlation problem is to specifically include the interelectronic distance in the trial wavefunction. Another way is to use what is known as *configuration interaction*. This involves using a wavefunction that is a linear combination of functions representing different electronic configurations of the molecule — the linear coefficients can then be used as variation parameters. As an example, we can again consider the hydrogen molecule. The simple MO function for the ground state leads to the configuration $(1\sigma_g)^2$. This configuration has symmetry given by the term symbol $^1\Sigma_g^+$. The simplest configuration interaction treatment combines this configuration with the doubly-excited configuration $(1\sigma_u)^2$ which also has symmetry given by $^1\Sigma_g^+$. This leads to a variation function of the following type:

$$\Psi = c_1(1\sigma_g)^2 + c_2(1\sigma_u)^2 \tag{10-53}$$

We shall not prove it, but in general, only configurations with the same multiplicity and belonging to the same irreducible representation can be used in the configuration interaction wavefunction. It is for this reason that we have not included in Eq. (10-53) the singly-excited configuration $(1\sigma_g)(1\sigma_u)$, because it has the wrong symmetry. However, the function in Eq. (10-53) uses only those configurations arising from the simplest LCAO treatment, i.e., linear combination of the $1s$ atomic orbitals. It is possible to construct other molecular-orbitals by linear combination of other atomic orbitals, $2s$, $2p$, etc., as we have seen in the last section. These lead to other excited configurations of the correct symmetry that can be used in the configuration interaction calculation.

## SUMMARY

1. In this chapter we examined one of the major methods for the theoretical treatment of chemical bonding—*molecular-orbital (MO) theory*.

2. We started by considering the most generally used method for constructing molecular orbitals, the *linear combination of atomic orbitals (LCAO) method*. In this method we assume that molecular orbital functions can be determined by taking a linear combination of the atomic orbitals of the atoms that form the molecule. The energies are determined by way of a set of *secular equations*:

$$\sum_i a_i (H_{ki} - ES_{ki}) = 0 \tag{10-18}$$

$H_{ki}$ are the Hamiltonian integrals between the atomic orbitals; $a_i$ are the coefficients in the linear combination; $S_{ki}$ are the *overlap integrals* for the atomic orbitals.

3. We applied the LCAO method to the simplest molecule, the $H_2^+$ ion. Using the $1s$ atomic orbitals of the hydrogen atoms, we obtained two molecular orbitals, one bonding, the other antibonding; we labeled these $\psi_g$ and $\psi_u$ to show their symmetry with respect to inversion through the center of symmetry of the molecule.

4. We extended the LCAO treatment to the hydrogen molecule and other diatomic molecules.

5. Molecular orbitals for diatomic molecules are labeled according to the value for $\lambda$, which is the quantum number associated with the component of orbital angular momentum in the direction of the molecular axis, by the symmetry with respect to inversion through the center of symmetry, and by a number that gives the order of increasing energy for orbitals with the same value for $\lambda$ and the same inversion symmetry. There are other methods for labeling molecular orbitals.

6. The electronic configurations of diatomic molecules are determined by putting the electrons into the available molecular orbitals of lowest energy.

7. We concluded the chapter by very briefly mentioning some improvements that can be made to the simple MO theory — these include application of the Hartree-Fock SCF theory and *configuration interaction*.

## EXERCISES

**10-1** Solve Eq. (10-29) and thereby verify that Eqs. (10-35) and (10-36) are the correct expressions for the orbital energies of the hydrogen molecule ion.

**10-2** By substituting the energies given in Eqs. (10-35) and (10-36) in Eq. (10-18), verify the forms of the molecular orbitals for the hydrogen molecule ion, as given by Eqs. (10-37) and (10-38). (Hint: Remember that the factors $(2+2S)^{-1/2}$ and $(2-2S)^{-1/2}$ are determined by normalizing the functions obtained by substitution of the energies into Eq. (10-18).)

**10-3** Verify that Eq. (10-40) is the correct form for the overlap integral $S$. (Hint: Evaluate the integral in Eq. (10-39) using elliptical coordinates. The volume element for this coordinate system is given by

$$d\tau = \frac{R^3}{8}(\xi^2 - \eta^2)d\xi\, d\eta\, d\phi$$

The limits of integration are determined by: $1 \leqslant \xi \leqslant \infty, -1 \leqslant \eta \leqslant 1, 0 \leqslant \phi \leqslant 2\pi$. You will need to use the integral

$$\int_x^\infty x^n \exp(-ax)dx = \frac{n!\, e^{-ax}}{a^{n+1}}\left[1 + ax + \frac{1}{2!}(ax)^2 + \cdots + \frac{1}{n!}(ax)^n\right]$$

**10-4** Apply the separation of variables method to Eq. (10-47) and show that $\Psi = \psi_g(1)\psi_g(2)$, where $\psi_g$ is the $H_2^+$ molecular orbital function, is a solution of the equation.

**10-5**   Using MO theory, explain the following facts: (a) the binding energy of $N_2^+$ is less than the binding energy of $N_2$; (b) the binding energy of $O_2^+$ is greater than the binding energy of $O_2$.

**10-6**   Examine the electronic configurations of the homonuclear diatomic molecules formed from the elements of the first row of the periodic table, and thereby verify (apart from the $+$ and $-$ designations) the ground state term symbols given in Table 10-1.

# 11

# Valence-bond Theory

## 11-1 INTRODUCTION

In the last chapter we examined MO theory, one of the major theoretical methods for dealing with chemical bonding; in this chapter we examine the other major method — *valence-bond (VB) theory*. We shall not spend very much time on VB theory because it is not used as much as MO theory in modern calculations. This should not imply that the latter theory is better. The mathematical difficulties inherent in VB theory discourage its use, but VB wavefunctions may be more accurate than MO functions. Regardless of all this, VB theory is important from the point of view of the aid it provides in our formulation of the general picture of chemical bonding, one of the major problems of quantum chemistry. In fact, it can be considered that the first real success of quantum chemistry came in 1927 when Heitler and London[13] managed to explain the bonding in the hydrogen molecule — the ideas introduced by them form the basis of VB theory.

In this chapter we shall first describe and examine the Heitler-London (HL) treatment of the hydrogen molecule. Then we shall briefly consider how the theory can be extended to more complex molecules. We shall compare the MO and VB treatments of the hydrogen molecule, and indicate how the HL wavefunction can be improved. Finally, we shall examine two concepts that are commonly used in VB theory — resonance and hybridization of atomic orbitals.

## 11-2 THE HYDROGEN MOLECULE

In our MO treatment of the hydrogen molecule the nuclei were first assigned to the relative positions they occupy in the molecule; then the electrons were assigned to the molecular orbitals formed by linear combination of the atomic orbitals of the individual hydrogen atoms. The Heitler-London approach to the problem is basically different. In this method the atoms are brought together in total and allowed to react.

Consider a system of two hydrogen atoms $A$ and $B$ in their ground states. If the atoms are far enough apart so that they do not react, each will have its electron in a $1s$ orbital, so we can write the wavefunctions for the electrons as $1s_A(1)$ and $1s_B(2)$. If we now bring the atoms together to form a molecule, one possibility for the total wavefunction for the system would be $1s_A(1)1s_B(2)$ — here we are assuming that electron 1 is associated with nucleus $A$ and electron 2 with nucleus $B$. However, such a wavefunction does not allow for the indistinguishability of the electrons, a subject we dealt with in some detail in Section 9-6. Because the electrons are indistinguishable, the wavefunction $1s_A(2)1s_B(1)$ is just as acceptable as the first one we considered. This led Heitler and London to choose a linear combination of the two functions. The function they used was

$$\Psi = (2+2S^2)^{-1/2}[1s_A(1)1s_B(2)+1s_A(2)1s_B(1)] \qquad (11\text{-}1)$$

where the factor before the bracket is a normalization constant that includes the overlap integral $S$.

The function in Eq. (11-1) does not take account of the spin of the electrons. However, we shall ignore this for the moment and proceed to examine the energy corresponding to this function. Later, when we examine the complete wavefunctions for the system, we shall have to consider electron spin. (The spin does not, in any case, contribute to the energy since we shall be using a spin-free Hamiltonian.) The energy corresponding to Eq. (11-1) is given by

$$E = \int \Psi \hat{H} \Psi d\tau \qquad (11\text{-}2)$$

The Hamiltonian is (cf. Fig. 10-6)

$$\hat{H} = -\frac{1}{2}(\nabla_1^2 + \nabla_2^2) - \frac{1}{r_{A1}} - \frac{1}{r_{B2}} - \frac{1}{r_{A2}} - \frac{1}{r_{B1}} + \frac{1}{r_{12}} + \frac{1}{R} \qquad (11\text{-}3)$$

With this Hamiltonian substituted in Eq. (11-2), (Exercise 11-2), we get

$$E = 2E_{\mathrm{H}} + \frac{Q+A}{1+S^2} \qquad (11\text{-}4)$$

Here $E_{\mathrm{H}}$ is the energy of the hydrogen atom in its ground state; $Q$ and

*A* are given by

$$Q = -\int 1s_A(1)\left(\frac{1}{r_{B1}}\right)1s_A(1)d\tau - \int 1s_B(2)\left(\frac{1}{r_{A2}}\right)1s_B(2)d\tau$$

$$+ \int 1s_A(1)1s_B(2)\left(\frac{1}{r_{12}}\right)1s_A(1)\,1s_B(2)d\tau + \frac{1}{R} \qquad (11\text{-}5)$$

$$A = -S\int 1s_A(1)\left(\frac{1}{r_{B1}}\right)1s_B(1)d\tau - S\int 1s_B(2)\left(\frac{1}{r_{A2}}\right)1s_A(2)d\tau$$

$$+ \int 1s_A(1)\,1s_B(2)\left(\frac{1}{r_{12}}\right)1s_A(2)\,1s_B(1)d\tau + \frac{S^2}{R} \qquad (11\text{-}6)$$

There are some observations we can make about this result:

1. $2E_H$ is the energy of two separated hydrogen atoms. Therefore, we can regard the second term in Eq. (11-4) as the energy change brought about by the interaction of the atoms, i.e., by the formation of the molecule.

2. $Q$ is a coulomb term because it is just the classical energy due to the electrostatic interactions between the two atoms — the first and second integrals represent the attractions of electron densities $(1s_A)^2$ to nucleus $B$ and of $(1s_B)^2$ to nucleus $A$; the third takes account of the repulsion between the two electron densities $(1s_A)^2$ and $(1s_B)^2$; the last accounts for the nuclear repulsion. The contribution to the binding energy of the molecule due to $Q$ is that which would result if we used only one of the functions $1s_A(1)1s_B(2)$ or $1s_A(2)1s_B(1)$; i.e., if we did not allow for electron exchange. This contribution is very small — less than ten percent of the experimental binding energy.

3. $A$ is an exchange term. A glance at the integrals in it shows us immediately that it is a result of our allowing for electron exchange, i.e., for the indistinguishability of the electrons. It is this term that accounts for the major portion of the binding energy calculated using Eq. (11-1) to describe the system.

4. The dissociation energy calculated from Eq. (11-4) is 0.116 hartrees. The experimental value is 0.174 hartrees. This method of dealing with the bonding in the hydrogen molecule, like the MO method we examined in Section 10-6, does therefore explain a large proportion of the binding energy.

We shall examine these results in a little more detail in Section 11-4.

But before we do this we shall look at some further aspects of the approach we have taken here.

The wavefunction in Eq. (11-1) resulted from taking a linear combination of the functions $1s_A(1)1s_B(2)$ and $1s_A(2)1s_B(1)$ in which we allowed for the indistinguishability of the electrons. What result would we have obtained if we had used the following function?

$$\Psi = (2 - 2S^2)^{-1/2}[1s_A(1)\,1s_B(2) - 1s_A(2)1s_B(1)] \qquad (11\text{-}7)$$

(Again, the factor before the bracket is a normalization constant.) This function also allows for the indistinguishability of the electrons. If we calculated the energy using this function, we would determine it to be given by

$$E = 2E_H + \frac{Q - A}{1 - S^2} \qquad (11\text{-}8)$$

If we actually evaluated this energy, we would discover it to be higher than the energy given by Eq. (11-4). Therefore, the original function we chose corresponds to the ground state of the molecule. In fact, the function Eq. (11-7) describes a repulsive state of the system.

Neither Eq. (11-1) nor Eq. (11-7) takes account of the spin of the electrons. To determine the complete wavefunction we must in fact include the spin. For two electrons we can write the following four spin functions:

$$\alpha(1)\,\beta(2) - \alpha(2)\,\beta(1)$$

$$\alpha(1)\,\alpha(2)$$

$$\beta(1)\,\beta(2) \qquad (11\text{-}9)$$

$$\alpha(1)\,\beta(2) + \alpha(2)\,\beta(1)$$

We must therefore combine these spin functions with the spatial functions Eqs. (11-1) and (11-7). However, we cannot use all of the possible combinations because the complete functions must be antisymmetric to exchange of the coordinates of the two electrons. With this restriction we can only combine Eq. (11-1) with the first of the functions in Eq. (11-9): Eq. (11-1) is a symmetric function, so to obtain a complete wavefunction that is antisymmetric we must combine it with an antisymmetric spin function — only the first function in Eq. (11-9) is antisymmetric. On the other hand, Eq. (11-7) is antisymmetric, so we can combine it with any of the last three functions in Eq. (11-9), all of which are symmetric. The complete functions are:

$$(2+2S^2)^{-1/2}[1s_A(1)1s_B(2)+1s_A(2)1s_B(1)]\frac{1}{\sqrt{2}}[\alpha(1)\beta(2)-\alpha(2)\beta(1)]$$

$$(2-2S^2)^{-1/2}[1s_A(1)1s_B(2)-1s_A(2)1s_B(1)]\begin{bmatrix}\alpha(1)\alpha(2)\\\beta(1)\beta(2)\\\frac{1}{\sqrt{2}}[\alpha(1)\beta(2)+\alpha(2)\beta(1)]\end{bmatrix}$$

$$(11\text{-}10)$$

The factor $1/\sqrt{2}$ is a normalization constant for the spin functions. The first function represents a state in which the spins of the two electrons are opposed; the total spin is zero, and the function therefore represents a singlet state $(2S+1=1)$. This is the function that corresponds to the electron-pair bond that accounts for the stability of the molecule. The other three functions represent a triplet state since the total spin is one $(2S+1=3)$ – this is the repulsive state, and in the absence of external fields the three components have the same energy.

## 11-3  EXTENSION OF THE HEITLER-LONDON TREATMENT

The Heitler-London treatment of the hydrogen molecule can be extended to more complex molecules. This was done by Slater[14] and Pauling[15]. The general theory is based on the concept of the electron-pair bond, and the hydrogen molecule is, of course, the simplest example. In general, if we consider two atoms $A$ and $B$ containing orbitals $\phi_A$ and $\phi_B$, we can write an antisymmetrized function $\Psi$ to represent the electron-pair bond formed by these two orbitals; in analogy with Eq. (11-10), it is

$$\Psi = N[\phi_A(1)\phi_B(2)+\phi_A(2)\phi_B(1)]\frac{1}{\sqrt{2}}[\alpha(1)\beta(2)-\alpha(2)\beta(1)]$$

$$(11\text{-}11)$$

where $N$ is a normalization constant. We can refer to this function as a *bond function*; it contains two electrons that have opposing spins. It is convenient to write this function in the following form:

$$\Psi = N[|\phi_A\bar{\phi}_B|-|\bar{\phi}_A\phi_B|]\tag{11-12}$$

In this notation spatial orbitals with bars are to be combined with $\beta$ spin functions; those without bars are to be combined with $\alpha$ spin functions. The determinants are Slater determinants expressed in the notation we introduced in Section 9-7 in which only the diagonal elements are written. To clarify this notation, let us write out one of the determinants in Eq. (11-12) in full:

$$|\phi_A\overline{\phi_B}| = \frac{1}{\sqrt{2}} \begin{vmatrix} \phi_A\alpha(1) & \phi_B\beta(1) \\ \phi_A\alpha(2) & \phi_B\beta(2) \end{vmatrix}$$

$$= \frac{1}{\sqrt{2}} [\phi_A\alpha(1)\phi_B\beta(2) - \phi_A\alpha(2)\phi_B\beta(1)]$$

In order to illustrate what we have just said, let us indicate very briefly how we would deal with a molecule more complicated than hydrogen — we choose the nitrogen molecule as an example. In the separated atoms each atom has three electrons in $2p$ atomic orbitals, with one electron per orbital. (We assume that the $1s$ and $2s$ electrons are non-bonding.) We can expect therefore that three bonds will form. We can consider that one bond will form between the orbitals that project along the internuclear axis (the $p_z$ orbitals) and two others between the orbitals that project perpendicularly to this axis (the $p_x$ and $p_y$ orbitals). If we label the atoms $A$ and $B$, using Eq. (11-12), we can write the bond functions

$$\Psi_1 = [|p_{zA}\overline{p_{zB}}| - |\overline{p_{zA}}p_{zB}|]$$

$$\Psi_2 = [|p_{xA}\overline{p_{xB}}| - |\overline{p_{xA}}p_{xB}|] \qquad (11\text{-}13)$$

$$\Psi_3 = [|p_{yA}\overline{p_{yB}}| - |\overline{p_{yA}}p_{yB}|]$$

To obtain the complete function for the bonding in the molecule we must combine the bond functions in a way that ensures that the indistinguishability of the electrons is allowed for. We shall not go into the details of doing this, but the function turns out to be a linear combination of $6 \times 6$ determinants; it is as follows:

$$\begin{aligned} \{&|p_{zA}\overline{p_{zB}}p_{xA}\overline{p_{xB}}p_{yA}\overline{p_{yB}}| - |\overline{p_{zA}}p_{zB}p_{xA}\overline{p_{xB}}p_{yA}\overline{p_{yB}}| \\ &- |p_{zA}\overline{p_{zB}}\,\overline{p_{xA}}p_{xB}p_{yA}\overline{p_{yB}}| - |p_{zA}\overline{p_{zB}}p_{xA}\overline{p_{xB}}\overline{p_{yA}}p_{yB}| \\ &+ |\overline{p_{zA}}p_{zB}\overline{p_{xA}}p_{xB}p_{yA}\overline{p_{yB}}| + |\overline{p_{zA}}p_{zB}p_{xA}\overline{p_{xB}}\overline{p_{yA}}p_{yB}| \\ &+ |p_{zA}\overline{p_{zB}}\overline{p_{xA}}p_{xB}\overline{p_{yA}}p_{yB}| - |\overline{p_{zA}}p_{zB}\overline{p_{xA}}p_{xB}\overline{p_{yA}}p_{yB}|\} \end{aligned} \qquad (11\text{-}14)$$

We have carried this discussion far enough to show that the VB function is rather complicated; it is more complicated, and consequently less easy to manipulate, than the simple, single-configuration MO function for the molecule. This is a general disadvantage of VB theory. And to improve the simple VB functions we have been discussing we have to introduce further complications – in the next sections we briefly indicate some of the improvements that can be made.

## 11-4 COMPARISON OF VB AND MO THEORY FOR HYDROGEN

Simple MO theory (cf. Section 10-6) gave a dissociation energy for the hydrogen molecule of 0.097 hartrees. Simple VB theory gives a value of 0.116 hartrees – this is somewhat nearer the experimental value. However, we should not conclude from this that VB theory is better than MO theory; both theories, in their simple forms, give results that are so far from the correct ones that there is little point in trying to compare them. At the same time we should not forget that both account for a considerable portion of the stability of the hydrogen molecule.

How can we improve the simple HL treatment of the hydrogen molecule? Let us first compare the simple MO wavefunction with the HL function. The spatial function for the HL treatment is given by Eq. (11-1):

$$\Psi_{HL} = [1s_A(1)\,1s_B(2) + 1s_A(2)\,1s_B(1)] \qquad (11\text{-}15)$$

(We have omitted the normalization constant since it is not important in this discussion.) The comparable MO function is (cf. Eq. (10-51))

$$\Psi_{MO} = [1s_A(1)\,1s_A(2) + 1s_B(1)\,1s_B(2) + 1s_A(1)\,1s_B(2) + 1s_A(2)\,1s_B(1)] \qquad (11\text{-}16)$$

In Section 10-8 we discussed the MO function and concluded that the first two terms can be considered to be ionic terms, representing the structures $H_A^- H_B^+$ and $H_A^+ H_B^-$; the last two terms can be considered to be covalent terms. Furthermore, the MO function gives equal weight to the ionic and covalent terms. At large internuclear distances this is not a a satisfactory function – when a quantity of hydrogen molecules dissociate, it suggests that half should give atoms and the other half $H^+$ and $H^-$ ions; in fact, the dissociation product is only atoms.

Turning to the HL function, we see that it contains only covalent terms. This function gives a somewhat better value for the binding energy than the MO function. It therefore seems that it is better to use a function with no ionic terms than one that gives very great weight to ionic terms.

We can get a better value for the energy than either the simple MO or simple VB energies by adding ionic terms to the HL function and using the variation theorem. This type of function is called a *Weinbaum func-*

*tion* [16]; the spatial part of it is

$$[\{1s_A(1)\,1s_B(2) + 1s_B(1)\,1s_A(2)\} + c\{1s_A(1)\,1s_A(2) + 1s_B(1)\,1s_B(2)\}]$$

(11-17)

$c$ is a variation parameter that can be adjusted to give the best value for the energy. A function of this type gives a slight improvement in the binding energy. However, if instead of using the regular $1s$ atomic orbitals of hydrogen to form the function we use atomic orbitals that include a variation parameter, we get considerable improvement in the binding energy. This is known as *scaling* the function — it is a general method for obtaining improved functions. In this case we could introduce a scale factor into the exponent of the $1s$ orbitals. This is equivalent to the scaling that we discussed in some detail with regard to the helium atom (Section 9-4) and STO's (Section 9-13) when we used the effective nuclear charge as a scale factor in the exponents of the atomic orbitals. The scaled Weinbaum function gives a dissociation energy of 0.1479 hartrees — this is quite an improvement on the previous functions we have discussed.

We should compare what we have done here with the method we used to improve the MO function, i.e., configuration interaction (cf. Section 10-8). Using just the one excited configuration $(1\sigma_u)^2$, configuration interaction gave a function of the following type:

$$\Psi = c_1(1\sigma_g)^2 + c_2(1\sigma_u)^2 \qquad (11\text{-}18)$$

If we expand Eq. (11-18), the spatial part of it has the form

$$a_1[1s_A(1)1s_B(2) + 1s_B(1)1s_A(2)] + a_2[1s_A(1)1s_A(2) + 1s_B(1)1s_B(2)]$$

(11-19)

where $a_1$ and $a_2$ are factors containing $c_1$ and $c_2$. Obviously, this function is of the same type as Eq. (11-17). Thus the addition of ionic terms into the VB function gives essentially the same result as using configuration interaction in MO theory. In fact, this is generally true for molecular bonding calculations: if the same set of basis orbitals is used, MO and VB theories are equivalent provided all possible ionic structures are included in the VB treatment and all possible electronic configurations are included in the MO treatment.

Another method for improving the HL function is based on recognition of the fact that when the two atoms approach each other the electron densities cannot be expected to remain symmetrical about the nuclei.

These densities will in fact be polarized towards the opposite nuclei. We might expect, therefore, that the spherically symmetrical $1s$ atomic orbitals are not good basis orbitals for construction of the VB function. We can allow for polarization by adding some $2p$ character to the $1s$ orbitals, and using the hybrid orbitals that result as basis orbitals for formation of the VB function. The form of these hybrid orbitals will be:

$$\phi = 1s + k2p_z \qquad (11\text{-}20)$$

(We are assuming that the internuclear axis is the $z$-axis.) $k$ is a variation parameter. Further improvement can be obtained by scaling the $1s$ and $2p$ orbitals. Rosen[17] used a function of this type and calculated a value for the dissociation energy about the same as that given with the Weinbaum function.

We should just mention before leaving this section that the improved values quoted above for the dissociation energy of the hydrogen molecule are certainly not the most accurate that can be calculated. In fact, modern electronic computers allow extremely accurate determination of the energy. For example, Kolos and Roothaan[18] and Kolos and Wolniewicz [19], using functions with very large numbers of terms have calculated energies that for all practical purposes are the same as the experimental value.

## 11-5 HYBRIDIZATION OF ORBITALS AND RESONANCE

To conclude this chapter we should say something about two concepts that are commonly used in VB theory — *hybridization of atomic orbitals* and *resonance*. No doubt these are familiar ideas to you, so we shall just mention them here.

VB theory is based on the concept of the electron-pair bond. In a molecule, electron-pair bonds form where there are large overlaps between atomic orbitals of the atoms that make up the molecule; further, we can associate a direction with a bond as that direction in which the orbitals forming the bond are concentrated. These qualitative ideas are dealt with in VB theory by invoking hybridization of orbitals when this would lead to stronger bonding.

Consider methane, $CH_4$, for example. The carbon atom has a triplet ground state ($^3P$), which means that it has two unpaired electrons — these are in $p$ orbitals. We would expect carbon to be divalent. In methane it is tetravalent. We explain this by saying that the carbon atom forms four $sp^3$ hybrid orbitals from its one $2s$ and three $2p$ orbitals. These are directed towards the corners of a tetrahedron, thus accounting for the tetrahedral shape of methane: the $1s$ orbitals of the four hydrogen atoms form large overlaps with the $sp^3$ orbitals of the carbon, thus accounting for strong bonds. We can write the hybrid orbitals very simply as

$$\phi_1 = \tfrac{1}{2}(s + p_x + p_y + p_z)$$
$$\phi_2 = \tfrac{1}{2}(s + p_x - p_y - p_z)$$
$$\phi_3 = \tfrac{1}{2}(s - p_x + p_y - p_z)$$
$$\phi_4 = \tfrac{1}{2}(s - p_x - p_y + p_z)$$

(11-21)

It is these orbitals that form the electron-pair bonds with the $1s$ hydrogen orbitals rather than the original $s$ and $p$ atomic orbitals.

There are many other examples of hybridization of orbitals we could quote. The important point to note here however is that VB theory uses the hybrid orbitals as basis orbitals to build up molecular functions. Hybridization can be used in MO theory, but then it is just a matter of convenience: the bonding orbitals in MO theory are molecular orbitals, and it does not matter whether these are constructed from the original atomic orbitals or from hybrid atomic orbitals — the result is the same.

Resonance plays an important part in VB theory. When examining the hydrogen molecule, we saw that the addition of ionic-type terms to the VB function (cf. Eq. (11-17)) improved the calculation of the energy. For that example we can consider that the covalent $(H-H)$ and ionic $(H^+ - H^-$ and $H^- - H^+)$ structures are *resonance structures*. In general, if several possible VB structures exist for a given molecule, then all have to be included in the VB function. The relative importance of the structures will depend on their relative energies. If one structure gives a much lower energy than the others, it might be satisfactory to ignore the others. For example, only one ionic structure is important for the HF molecule, and we would write the wavefunction:

$$\Psi(\mathrm{HF}) = c_1 \Psi(\mathrm{H{-}F}) + c_2 \Psi(\mathrm{H^+{-}F^-})$$

(11-22)

where $c_1$ and $c_2$ are variation parameters.

A word of warning before leaving the question of resonance: When we say that one or more resonance structures exist for a molecule we do not mean that the molecule is jumping back and forth between the structures. The resonance structures have no real existence — resonance is merely a convenient device for picturing the mathematical concepts.

## SUMMARY

1. In this chapter we briefly examined *valence-bond (VB) theory* and compared it to MO theory.

2. We examined the Heitler-London treatment of the hydrogen molecule. The wavefunction used in this treatment is as follows:

$$\Psi = (2 + 2S^2)^{-1/2}[1s_A(1)1s_B(2) + 1s_A(2)1s_B(1)] \qquad (11\text{-}1)$$

This function takes account of the indistinguishability of the electrons.

3. We discovered that the bonding energy of the molecule calculated using the HL function is somewhat better than that given by simple MO theory. The major part of the calculated binding energy is due to the presence of the exchange term in the energy expression, i.e., the term resulting from the electrons' indistinguishability.

4. We briefly indicated how the HL treatment of the hydrogen molecule can be extended to more complex molecules. If two atoms $A$ and $B$ contain orbitals $\phi_A$ and $\phi_B$, we can represent the bond that forms from these orbitals by a bond function given by

$$\Psi = N[\phi_A(1)\phi_B(2) + \phi_A(2)\phi_B(1)] \frac{1}{\sqrt{2}} [\alpha(1)\beta(2) - \alpha(2)\beta(1)] \quad (11\text{-}11)$$

This function represents an electron-pair bond in which the electrons have opposing spins. There is one such function for each electron-pair bond in the molecule.

5. We compared the MO and VB treatments for the hydrogen molecule and saw that by including ionic terms in the VB function, the VB method could be made to give essentially the same result as the MO method using configuration interaction.

6. We concluded the chapter with a brief examination of *resonance* and *hybridization* of the atomic orbitals for use in construction of VB functions.

## EXERCISES

**11-1** Prove that the normalization constant in Eq. (11-1) is correct.

**11-2** Using the wavefunctions in Eqs. (11-1) and (11-7) and the Hamiltonian in Eq. (11-3), show that the energy expressions Eqs. (11-4) and (11-8) are correct.

**11-3** Show that the wavefunctions in Eqs. (11-11) and (11-12) are equivalent.

**11-4**  Using the $2s$ and $2p_z$ (taking $z$ as the internuclear axis) orbitals of lithium, write down the VB bond functions for LiH that are analogous to those for $N_2$ given in Eq. (11-13). Suggest how this simple VB treatment of the LiH molecule might be improved.

**11-5**  Show that the hybrid orbitals in Eq. (11-21) are mutually orthogonal.

# 12

# Hückel Molecular-orbital Theory

## 12-1  INTRODUCTION

In this and the next chapters we shall very briefly indicate how quantum chemistry is applied to systems of real chemical interest. The systems we have been discussing in some detail—the hydrogen atom, the helium atom, the hydrogen molecule, etc.—are certainly of interest to chemists, but, after all, the vast majority of problems of chemical interest involve far larger and far more complex molecules than these. It is not possible to solve the Schroedinger equation exactly for large molecules. This does not mean however that quantum mechanics has nothing to offer when we wish to deal with large molecules—on the contrary. In this chapter we shall illustrate how even gross approximations to exact theory can sometimes give useful results—we are going to examine an application of MO theory to larger molecules than we have so far discussed; the theory is known as *Hückel molecular-orbital (HMO) theory*[20], and it has been very useful to organic chemists for the correlation of properties of unsaturated molecules.

First we shall examine the basic approximations needed for application of the theory. Then we shall see how the theory is applied to some very simple systems.

## 12-2  HÜCKEL MO THEORY

A very important class of unsaturated organic molecules includes those molecules that have *conjugated* double bonds. These molecules have classical formulae that include alternating single and double C–C

210

bonds. Examples are: butadiene, benzene, naphthalene, and the purine and pyrimidine bases of DNA. In fact, a very large number of important organic and biochemical molecules are of this type. Many of these molecules have been studied in great detail by experimentalists, resulting in a vast accumulation of experimental data. It has become very desirable, therefore, to try to understand the experimental results for these molecules in terms of quantum theory: Hückel molecular-orbital theory has made a major contribution in this regard.

The distinctive feature of conjugated molecules is that we can classify their valence electrons into two separate sets, $\pi$-electrons and $\sigma$-electrons. The $\sigma$-electrons are assumed to be strongly localized in the individual bonds, the bond orbitals associated with them having the $\sigma$-type symmetry we discussed for diatomic molecules (cf. Section 10-7). These electrons are relatively unreactive. On the other hand, the $\pi$-electrons are relatively much more reactive and therefore play a more important role in the chemical reactions of the molecules. $\pi$-electrons, unlike the $\sigma$-electrons, are assumed to be delocalized over the carbon framework of the molecule. Their name derives from the analogy between the molecular orbitals describing them and the $\pi$-type molecular orbitals we discussed for diatomic molecules. The classic example of a $\pi$-electron system is, of course, that which occurs in the benzene molecule; in this molecule we assume that six $p$ orbitals, one from each of the carbon atoms, overlap to form the $\pi$-orbital system—the six $\pi$-electrons are then delocalized over the whole molecule when we assign them to this system. It is $\pi$-electron systems of this type that we are going to be concerned with here.

We first assume that the wavefunction for the $\pi$-electrons is independent of the $\sigma$-electron framework; i.e., the $\pi$-electron system acts independently. This is a rather drastic assumption, but it can be rationalized in terms of experimental results for conjugated organic molecules, which seem to suggest that the $\pi$-electron system does act independently and accounts for the characteristic properties of conjugated molecules. We also justify the assumption in terms of the results it gives—in fact, this is how we justify all the assumptions of Hückel molecular-orbital theory.

Even after having made this assumption, we are still in trouble: the Hamiltonian for the $\pi$-electron system is too complicated to allow us to find exact solutions for the equation that gives the energies of the $\pi$-electrons. For example, there are electron-electron interaction terms in it, and we have previously seen that these make it impossible to obtain exact solutions. We assume, therefore, that the Hamiltonian for the $\pi$-electron system can be written as a sum of one-electron terms:

$$\hat{H} = \sum_i \hat{h}_{\text{effective}}(i) \tag{12-1}$$

$i$ refers only to the $\pi$-electrons. $\hat{h}_{\text{effective}}(i)$ is a one-electron operator that is a function only of the coordinates and momenta of electron $i$—this operator includes in its potential energy part an allowance for the effect of the other electrons on electron $i$. Since $\hat{h}_{\text{effective}}(i)$ is a function only of the coordinates and momenta of electron $i$, we can use the separation of variables technique on the $\pi$-electron Schroedinger equation resulting from the Hamiltonian in Eq. (12-1). This gives us $N$ one-electron equations to solve, where $N$ is the number of $\pi$-electrons—each equation has the form

$$\hat{h}_{\text{effective}}(i)\psi_i = E_i\psi_i \tag{12-2}$$

The $\psi_i$ are one-electron molecular orbitals, and the $E_i$ are the corresponding orbital energies. The total $\pi$-electron energy is then given by

$$E = \sum_i E_i \tag{12-3}$$

When performing this summation, we must keep in mind which orbitals are occupied—the Pauli principle must be taken into account when we place the electrons in the orbitals.

We now apply the regular LCAO method for constructing molecular orbitals (cf. Section 10-4). That is, we assume that the molecular orbitals $\psi_i$ can be written as linear combinations of atomic orbitals:

$$\psi_i = \sum_\mu a_{\mu i}\phi_\mu \tag{12-4}$$

$\phi_\mu$ are the $p$ orbitals that contribute to the $\pi$-system, and the summation is over all the atoms in this system. Using the Hamiltonian in Eq. (12-1) and the variation theorem, we get a set of secular equations that have a non-trivial solution if the secular determinant is zero. We therefore get an equation of the form:

$$\begin{vmatrix} H_{11} - ES_{11} & H_{12} - ES_{12} & \cdots & H_{1N} - ES_{1N} \\ H_{21} - ES_{21} & H_{22} - ES_{22} & \cdots & H_{2N} - ES_{2N} \\ \cdots\cdots\cdots\cdots\cdots\cdots\cdots\cdots\cdots\cdots\cdots\cdots \\ H_{N1} - ES_{N1} & H_{N2} - ES_{N2} & \cdots & H_{NN} - ES_{NN} \end{vmatrix} = 0 \tag{12-5}$$

or more concisely,

$$|H_{\mu\nu} - ES_{\mu\nu}| = 0 \tag{12-6}$$

Here the basis set of orbitals are the atomic orbitals that contribute to the $\pi$-system.

HMO theory at this point makes some further gross approximations, which, as we have already mentioned, are justified by the results they give; these approximations are:

1. $H_{\mu\mu}$ is the same for each atom; it is usually given the symbol $\alpha$.
2. $H_{\mu\nu}$ is a constant if atoms $\mu$ and $\nu$ are bonded together; it is usually given the symbol $\beta$.
3. $H_{\mu\nu} = 0$ if atoms $\mu$ and $\nu$ are not bonded.
4. $S_{\mu\nu} = 1$ if $\mu = \nu$.
5. $S_{\mu\nu} = 0$ if $\mu \neq \nu$; i.e., there is zero overlap even between adjacent atomic orbitals.

We shall see in later sections of this chapter that when we use HMO theory, the results are expressed in terms of the energy parameters $\alpha$ and $\beta$. We should therefore mention at this point that we do not try to calculate these parameters. We have defined them by integrals that include an ill-defined Hamiltonian, Eq. (12-1). We get out of the difficulty of exactly defining this Hamiltonian by choosing empirical values for $\alpha$ and $\beta$ that are in accord with experimental results. For example, using the $\pi$-electron model, we could measure experimentally the energy that would correspond to excitation of a $\pi$-electron from the ground state to the first excited state; we could then correlate the result with that calculated by HMO theory. If we measured this transition energy for butadiene — to give a specific example — we would determine it to be 5.9 eV. In Section 12-4 we show that the HMO calculation leads to a value of $-1.24\beta$ — this allows calculation of an empirical value for $\beta$. Unfortunately, different values for $\beta$ are given by different experimentally determined properties. This, however, is not surprising considering the nature of the Hückel approximations. These approximations are basically quite naïve. In the butadiene case, for example, we have not taken account of the different carbon-carbon bond lengths — this difference obviously will affect the value of $\beta$, and we might expect that the use of two different values of $\beta$, one for the single bonds, the other for the double bonds, would be more realistic. Another situation where the value of $\beta$ (and $\alpha$) must be adjusted is in the case of molecules with heteroatoms in the conjugated system, e.g., pyridine. These can be dealt with by HMO theory, but it is necessary to use different values of $\alpha$ and $\beta$ for the heteroatoms than for the carbon atoms.

It is possible to make other improvements on the very simple theory.

For example, the simple theory does not make any specific reference to interelectronic interactions, though we assume that these are somehow or other averaged in the Hückel Hamiltonian. One improvement, therefore, would be to try to specifically include interelectronic interactions in some way, and this in fact has been done†. Here, however, we shall concern ourselves only with the very simple theory, and we shall now proceed to give some examples of its application.

## 12-3  ETHYLENE

The ethylene molecule has sixteen electrons. To perform even an approximate calculation for an electron system this large, would be a major undertaking. The Hückel approximations reduce the problem to a two-electron calculation, the two electrons being the $\pi$-electrons. In Fig. 12-1 we show the molecule with the $2p$ orbitals that are the basis set for the $\pi$-electron calculation.

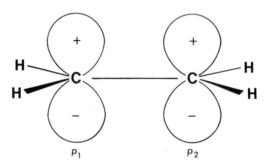

**Fig. 12-1**  The ethylene molecule showing the $2p$ atomic orbitals used in the Hückel calculation for the $\pi$-system.

Applying the LCAO method for the $\pi$-electrons, we get

$$\psi_1 = a_{1i}\phi_1 + a_{2i}\phi_2 \tag{12-7}$$

where $\phi_1$ and $\phi_2$ are the $2p$ atomic orbitals.

The secular determinant is a $2 \times 2$ determinant. To determine the energy we have to solve the following equation (cf. Eq. (12-5)):

$$\begin{vmatrix} H_{11} - ES_{11} & H_{12} - ES_{12} \\ H_{21} - ES_{21} & H_{22} - ES_{22} \end{vmatrix} = 0 \tag{12-8}$$

†For further details of adjustments to and improvements on the simple theory see, for example, reference B6 in the Bibliography, Chapter 15.

When we apply the Hückel approximations we get

$$\begin{vmatrix} \alpha - E & \beta \\ \beta & \alpha - E \end{vmatrix} = 0 \qquad (12\text{-}9)$$

It is normal to simplify the determinant by writing

$$x = \frac{(\alpha - E)}{\beta} \qquad (12\text{-}10)$$

This gives

$$\begin{vmatrix} x & 1 \\ 1 & x \end{vmatrix} = 0 \qquad (12\text{-}11)$$

i.e.,

$$x^2 - 1 = 0 \qquad (12\text{-}12)$$

$$x = \pm 1$$

Therefore, $E = \alpha \pm \beta$. Substituting these values back into the secular equations and normalizing in the usual way, we get two functions:

$$\psi_1 = \frac{1}{\sqrt{2}}(\phi_1 + \phi_2) \quad \text{with energy } \alpha + \beta \qquad (12\text{-}13\text{a})$$

$$\psi_2 = \frac{1}{\sqrt{2}}(\phi_1 - \phi_2) \quad \text{with energy } \alpha - \beta \qquad (12\text{-}13\text{b})$$

Since both $\alpha$ and $\beta$ are negative, $\psi_1$ is lower in energy than $\psi_2$. Therefore, both the $\pi$-electrons of the ethylene molecule in its ground state will be in $\psi_1$, and the total $\pi$-electron energy is $2(\alpha + \beta)$. Figure 12-2 is an energy level diagram for the $\pi$-electrons. In this diagram we have compared the orbital energies to $\alpha$. This latter quantity, remember, is given by $\int \phi_\mu \hat{H} \phi_\mu d\tau$ which can be regarded as the energy of a $p$-electron in an isolated carbon atom.

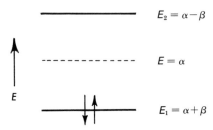

$$E_2 = \alpha - \beta$$

$$E = \alpha$$

$$E_1 = \alpha + \beta$$

**Fig. 12-2**  Energy level diagram for the Hückel calculation for the $\pi$-electrons of ethylene. The energy levels are shown relative to $\alpha$, the energy of a $p$-electron in an isolated carbon atom.

## 12-4 BUTADIENE

Though still a very simple system, butadiene is a somewhat more interesting example of the use of Hückel theory. We first apply the theory as we have done for ethylene. Later, in Section 12-8, we show how the use of the symmetry properties of the molecule can simplify the calculations.

We assume that the molecule is linear as shown in Fig. 12-3. (It actually exists in cis and trans forms, but simple Hückel theory does not show an

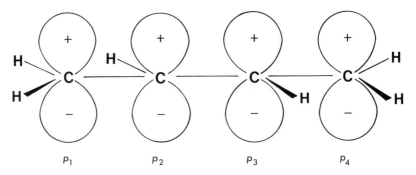

**Fig. 12-3** Schematic representation of the butadiene molecule with the p orbitals used in the Hückel calculation for the π-system. Note that for the Hückel calculation the molecule can be assumed to be linear; in reality it exists in cis and trans forms but HMO theory does not distinguish between the two.

energy difference between the forms—this is because it does not take into account the interactions between non-nearest neighboring atoms). There are four p orbitals in the basis set, so the secular equations give rise to a $4 \times 4$ determinantal equation; this is

$$\begin{vmatrix} x & 1 & 0 & 0 \\ 1 & x & 1 & 0 \\ 0 & 1 & x & 1 \\ 0 & 0 & 1 & x \end{vmatrix} = 0 \tag{12-14}$$

$x$ is given by Eq. (12-10). We solve this equation using the Laplace development of the determinant (cf. Section 8-4). Using the first row, we get

$$x \begin{vmatrix} x & 1 & 0 \\ 1 & x & 1 \\ 0 & 1 & x \end{vmatrix} - \begin{vmatrix} 1 & 1 & 0 \\ 0 & x & 1 \\ 0 & 1 & x \end{vmatrix} = 0$$

i.e.,

$$x^2 \begin{vmatrix} x & 1 \\ 1 & x \end{vmatrix} - x \begin{vmatrix} 1 & 1 \\ 0 & x \end{vmatrix} - \begin{vmatrix} x & 1 \\ 1 & x \end{vmatrix} = 0$$

$$x^4 - 3x^2 + 1 = 0$$

$$x = \pm 1.62, \pm 0.62 \tag{12-15}$$

Therefore, from Eq. (12-10) $E$ takes the values shown in Table 12-1. We show these values diagrammatically in Fig. 12-4, which is an energy

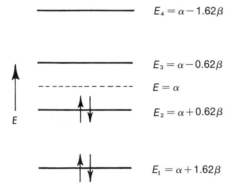

Fig. 12-4    Energy level diagram for the Hückel calculation for the $\pi$-electrons of butadiene. The energy levels are shown relative to $\alpha$.

level diagram for the $\pi$-electrons of butadiene. Again, as in the ethylene example, we have referred the energies to $\alpha$, the energy of a $p$-electron in an isolated carbon atom.

We can determine the coefficients for the $\psi$'s in

$$\psi_i = a_{1i}\phi_1 + a_{2i}\phi_2 + a_{3i}\phi_3 + a_{4i}\phi_4 \tag{12-16}$$

in the usual way. When we do this we get the values shown in Table 12-1.

**Table 12-1**    Hückel calculation results for butadiene.

| $\psi_i$ | $E_i$ | $a_{1i}$ | $a_{2i}$ | $a_{3i}$ | $a_{4i}$ |
|---|---|---|---|---|---|
| $\psi_1$ | $\alpha + 1.62\beta$ | 0.37 | 0.60 | 0.60 | 0.37 |
| $\psi_2$ | $\alpha + 0.62\beta$ | 0.60 | 0.37 | $-0.37$ | $-0.60$ |
| $\psi_3$ | $\alpha - 0.62\beta$ | 0.60 | $-0.37$ | $-0.37$ | 0.60 |
| $\psi_4$ | $\alpha - 1.62\beta$ | 0.37 | $-0.60$ | 0.60 | $-0.37$ |

$\psi_1$ and $\psi_2$ are bonding orbitals; $\psi_3$ and $\psi_4$ are antibonding. In the ground state the four $\pi$-electrons will be in $\psi_1$ and $\psi_2$. Therefore, the total $\pi$-electron energy is $4\alpha + 4.48\,\beta$. It is interesting to note that two isolated double bonds (ethylene-like bonds) would have an energy of $4\alpha + 4\beta$ (cf. Section 12-3). The two double bonds in butadiene have an energy $0.48\,\beta$ lower than this—we call this *delocalization energy*; it represents a stabilization due to the electrons being able to move over the whole molecule.

## 12-5 BENZENE

Since benzene is the classic example of a $\pi$-electron system, we shall very briefly outline the application of Hückel theory to this molecule. The details are very similar to those we have discussed for ethylene and butadiene, so we shall not repeat them. The results, however, are interesting insofar as they show how Hückel theory accounts for the large delocalization energy for the benzene ring system of $\pi$-electrons.

Because of the delocalization of the $\pi$-electrons, we cannot represent benzene with a single valence structure. In fact, if we insist on using valence structures to represent the molecule, we have to write several that are referred to as resonance structures. The two most important resonance structures are the Kekulé structures:

Other significant, but less important, resonance structures are the Dewar structures:

We can represent the delocalization of the $\pi$-electrons by superimposing these and other less significant resonance structures.

The Hückel treatment of benzene proceeds in exactly the same way as for the two examples we have already discussed in detail. The secular equation analogous to Eqs. (12-11) and (12-14) is

$$\begin{vmatrix} x & 1 & 0 & 0 & 0 & 1 \\ 1 & x & 1 & 0 & 0 & 0 \\ 0 & 1 & x & 1 & 0 & 0 \\ 0 & 0 & 1 & x & 1 & 0 \\ 0 & 0 & 0 & 1 & x & 1 \\ 1 & 0 & 0 & 0 & 1 & x \end{vmatrix} = 0 \qquad (12\text{-}17)$$

where again $x$ is given by Eq. (12-10). When expanded, this equation gives

$$x^6 - 6x^4 + 9x^2 - 4 = 0$$

This equation, in turn, leads to the following orbital energies:

$$\begin{aligned}
E_1 &= \alpha + 2\beta \\
E_2 &= E_3 = \alpha + \beta \\
E_4 &= E_5 = \alpha - \beta \\
E_6 &= \alpha - 2\beta
\end{aligned} \qquad (12\text{-}18)$$

The delocalization energy of benzene is the total $\pi$-electron energy minus the energy of the molecule if it were bonded with three ethylenic-type double bonds. Since we have six $\pi$-electrons, and since the orbitals of lowest energy are $E_1, E_2$ and $E_3$, we get:

delocalization energy $= 2(\alpha + 2\beta) + 4(\alpha + \beta) - 3(2\alpha + 2\beta) = 2\beta$

This delocalization energy is an additional stabilization of the molecule due to the delocalization of the $\pi$-electrons over the whole ring system. VB theory accounts for the stability of benzene in terms of resonance (cf. Section 11-5). In VB theory, we can define a *resonance energy* as the energy difference between the VB structure of lowest energy (one of the Kekulé structures in this case) and the true energy of the molecule, which would correspond to superposition of all the possible resonance structures. We might expect, therefore, that there would be some relationship between the delocalization energy of HMO theory and the resonance energy of VB theory; in fact, the two have been equated. It is possible to measure an experimental or *empirical resonance energy*. This can be done, for example, by measuring the difference between the actual heat of combustion of the molecule and the heat of combustion predicted on the basis of addition of the bond energies for the C—H, C—C and C=C bonds in a Kekulé type structure. If we calculate the delocalization energy for benzene determined by HMO theory and equate it to the empirical resonance energy measured in this way, we get a value for $\beta$ of about $-18$ kcal mole$^{-1}$. Actually, a large number of aromatic hydrocarbons, when treated on this basis, give a similar value for $\beta$. We have, however, already mentioned (cf. Section 12-2) that the values of $\beta$ given by different experimental methods differ considerably. This should alert us to the fact that there is more to this subject than simply equating the delocalization and resonance energies†.

---

†There is considerably more to this subject than we have discussed here. For further details you can consult reference F5 in the Bibliography, Chapter 9, or reference C2, pp.657–663.

## 12-6  HMO COEFFICIENTS AND $\pi$-ELECTRON DISTRIBUTION

One of the results given by the HMO treatment of a $\pi$-electron system is the coefficients for the molecular orbitals, i.e., the coefficients for the linear combination of atomic orbitals. Here we shall indicate how these coefficients can be used to give some information about the $\pi$-electron distribution in a molecule. This information is useful when we wish to examine theoretically the question of chemical reactivity of molecules with $\pi$-electron systems. We can use butadiene as an example for this discussion since we have previously given the coefficients for this molecule (cf. Table 12-1).

We define three quantities: $\pi$-electron charge density at a particular atom, $\pi$-electron bond order for a particular bond, and free-valence number.

1. An electron in the orbital $\psi_i = \sum_\mu a_{\mu i}\phi_\mu$ (cf. Eq. (12-4)) has associated with it a density distribution that can be given by $\sum_\mu a_{\mu i}^2\phi_\mu^2$ if we neglect all the overlap density terms of the type $\phi_\mu\phi_\nu$. We can therefore consider $a_{\mu i}^2$ to be a measure of the charge density due to this electron at atom $\mu$. By summing over all the electrons in all the various occupied molecular orbitals, we can define the $\pi$-*electron charge density* at the $\mu$th atom, $q_\mu$:

$$q_\mu = \sum_i n_i a_{\mu i}^2 \qquad (12\text{-}19)$$

$n_i$ is the number of electrons in the $i$th molecular orbital. For butadiene, we can use Table 12-1 and write: $q_1 = 2a_{11}^2 + 2a_{12}^2 = 2(0.37)^2 + 2(0.60)^2 = 1.00$; $q_2 = 2a_{21}^2 + 2a_{22}^2 = 2(0.60)^2 + 2(0.37)^2 = 1.00$. Because of the symmetry of the molecule, $q_3$ and $q_4$ are also equal to 1.00.

2. The second quantity we define is the $\pi$-*electron bond order* (also called the *mobile bond order*) of the bond between atoms $\mu$ and $\nu$. If atoms $\mu$ and $\nu$ are directly bonded, this is a measure of the multiple bond character of the bond. The bond order for the bond $\mu\nu$, $p_{\mu\nu}$, is defined by

$$p_{\mu\nu} = \sum_i n_i a_{\mu i} a_{\nu i} \qquad (12\text{-}20)$$

For butadiene: $p_{12} = 2a_{11}a_{21} + 2a_{12}a_{22} = 2(0.37 \times 0.60) + 2(0.60 \times 0.37) = 0.89$;    $p_{23} = 2a_{21}a_{31} + 2a_{22}a_{32} = 2(0.60 \times 0.60) + 2(0.37 \times -0.37) = 0.45$. Again, by symmetry, $p_{34} = p_{12} = 0.89$.

If we refer back to the ethylene calculation in Section 12-3, we see that the bond order for this molecule is one. According to HMO theory, therefore, the closer a bond order is to one, the greater the double bond character. The two end bonds in butadiene therefore have considerably greater double bond character than the central bond—this is in general accord with the normal representation of the butadiene molecule which shows the end bonds to be double and the central bond to be single.

Bond orders have been correlated with bond lengths and bond energies. In general, the larger the bond order, the shorter the bond length and the larger the bond energy.

3. The third quantity that is useful when discussing chemical reactivity is the so-called *free-valence number*. We define the free-valence number of atom $\mu$, $F_\mu$:

$$F_\mu = N_{max} - \sum_\nu p_{\mu\nu} \tag{12-21}$$

$N_{max}$ is the maximum value of the sum of the bond orders of all the bonds coming from atom $\mu$—this is usually taken to be $\sqrt{3}$, the maximum value for carbon in an $sp^2$ valence state; the summation is over all the bonds to atom $\mu$. For butadiene: $F_1 = F_4 = \sqrt{3} - 0.89 = 0.84$; $F_2 = F_3 = \sqrt{3} - 0.89 - 0.45 = 0.39$. The free-valence number can be approximately correlated with the susceptibility of an atom in a molecule to free-radical attack. The figures we have just calculated for butadiene are in accord with the known greater susceptibility to attack of the end carbon atoms.

The calculations we have just carried out for butadiene can be summarized by the following diagram:

$$CH_2 \text{————} CH \text{————} CH \text{————} CH_2$$

| $p$: | 0.89 | | 0.45 | | 0.89 | |
|---|---|---|---|---|---|---|
| $q$: 1.00 | | 1.00 | | 1.00 | | 1.00 |
| $F$: 0.84 | | 0.39 | | 0.39 | | 0.84 |

## 12-7  ALTERNANT HYDROCARBONS

The conjugated molecules we have chosen to illustrate HMO theory, butadiene and benzene, belong to a class of conjugated molecules known as *alternant hydrocarbons*. The definition of alternant hydrocarbons is as follows: these are molecules in which the carbon atoms can be divided into two sets such that the members of one set are bonded only to members of the other set. Figure 12-5 illustrates the meaning of this definition. It is usual, to refer to the one set of atoms as the starred set and to the other as the unstarred set—an alternant hydrocarbon is then one in which no two starred and no two unstarred atoms are bonded. In Fig. 12-5 we also show some examples of *non-alternant hydrocarbons*. You will notice that these non-alternant molecules are cyclic compounds with odd-membered rings—in general, such compounds are non-alternants.

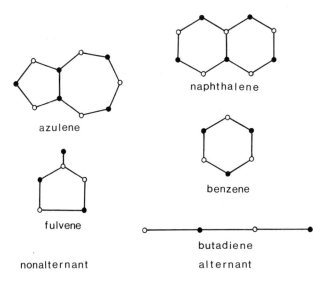

naphthalene

azulene

benzene

fulvene

butadiene

nonalternant    alternant

**Fig. 12-5**   The carbon frameworks of some alternant and non-alternant hydrocarbons. The starred atoms are shown as black circles, the unstarred as open circles.

The importance of distinguishing between alternant and non-alternant hydrocarbons is that HMO theory is more easily applied to the former than to the latter, and HMO theory results for non-alternant hydrocarbons are usually less satisfactory than for alternant molecules. Because of the special structural features of alternant hydrocarbons the mathematics involved in HMO calculations is relatively simpler than for non-alternants — this is indicated by what follows.

If we refer back to our HMO calculations for butadiene, we see that the orbital energies occur in pairs, $\alpha \pm 1.62\beta$ and $\alpha \pm 0.62\beta$. The orbital energies for benzene also occur in pairs. This is a general property of alternant hydrocarbons. An example that illustrates this property more dramatically is naphthalene. Naphthalene is an alternant hydrocarbon, and a HMO theory calculation of the orbital energies gives the values: $\alpha \pm 2.303\beta$, $\alpha \pm 1.618\beta$, $\alpha \pm 1.303\beta$, $\alpha \pm 1.000\beta$, $\alpha \pm 0.618\beta$. This pairing of energies is expressed by the *HMO pairing theorem:* For every HMO energy $\alpha + x\beta$ in an alternant hydrocarbon there is another HMO energy $\alpha - x\beta$. Furthermore, the coefficients of paired orbitals are either the same or merely change in sign. (A glance at Table 12-1 shows this latter statement to be true for butadiene.) The pairing properties of alternant hydrocarbons enable us to use no more than half the roots of the secular determinant to determine all the orbitals and orbital energies.

A special case of the pairing theorem occurs when the number of carbon atoms in the alternant molecule is odd, e.g., in the allyl radical $CH_2=CH-CH_2\cdot$. Then, one root is $x = 0$, i.e., $E = \alpha$, and the pairing is obviously trivial.

If we refer again to the calculations we performed on butadiene (Section 12-6), we see an illustration of another property that can be generalized. For alternant hydrocarbons that have $N$ $\pi$-electrons and $N$ carbon atoms the $\pi$-electron charge density at each carbon atom is unity. This means that these molecules should be non-polar. In fact, molecules like naphthalene and phenanthrene, which are alternant hydrocarbons, have no measurable dipole moments, whereas a non-alternant like azulene is found to be quite polar.

## 12-8  USE OF MOLECULAR SYMMETRY PROPERTIES IN HMO THEORY

In Chapter 8 we examined in some detail the methods whereby molecular symmetry properties can be used to simplify calculations. In that chapter we examined the application of group theory to molecular symmetry. Here we are not going to use group theory in a formal way, but we are going to examine how some of the ideas connected with molecular symmetry that we discussed in Chapter 8 can be used to simplify Hückel calculations.

HMO theory, as we have said before, is based on a number of serious approximations — these approximations however do allow us to apply MO theory to large molecules. Nevertheless, in many cases the application of HMO theory leads to very tedious computations. If the molecule possesses some symmetry, we can reduce the amount of computation. To illustrate this, we again choose butadiene as an example.

Though butadiene has other symmetry elements, we shall consider just the plane of symmetry $\sigma$ shown in Fig. 12-6. We can construct linear combinations of the $2p$ orbitals $\phi_\mu$ that are eigenfunctions of the symmetry operator $\hat{\sigma}$: they are

$$S_1 = \frac{1}{\sqrt{2}}(\phi_1 + \phi_4) \qquad S_3 = \frac{1}{\sqrt{2}}(\phi_2 - \phi_3)$$

$$S_2 = \frac{1}{\sqrt{2}}(\phi_2 + \phi_3) \qquad S_4 = \frac{1}{\sqrt{2}}(\phi_1 - \phi_4)$$

(12-22)

Let us now, instead of using the original $p$ orbitals, use these symmetry functions as a basis set for our secular equations. After applying the

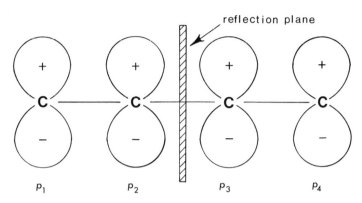

**Fig. 12-6**    The plane of symmetry for butadiene used in the discussion of Section 12-8. The hydrogen atoms of butadiene are not shown.

Hückel approximations, this allows us to write

$$
\begin{vmatrix}
H_{11}^s - E & H_{12}^s & H_{13}^s & H_{14}^s \\
H_{21}^s & H_{22}^s - E & H_{23}^s & H_{24}^s \\
H_{31}^s & H_{32}^s & H_{33}^s - E & H_{34}^s \\
H_{41}^s & H_{42}^s & H_{43}^s & H_{44}^s - E
\end{vmatrix} = 0 \qquad (12\text{-}23)
$$

We have put the superscript on the Hamiltonian integrals to emphasize that these are integrals over the symmetry orbitals, not over the original atomic orbitals; e.g., $H_{12}^s \equiv \int S_1 \hat{H} S_2 d\tau$.

Because the secular determinant in Eq. (12-23) is expressed in terms of the symmetry orbitals, we can now make use of an important quantum mechanical theorem to simplify it further. The theorem, which you can very easily prove (Exercise 12-3), says: If two operators $\hat{P}$ and $\hat{Q}$ commute, all integrals of the type $\int \psi_i^* \hat{Q} \psi_j d\tau$, where $\psi_i$ and $\psi_j$ are eigenfunctions of $\hat{P}$, are zero unless $\psi_i$ and $\psi_j$ have the same eigenvalue. We can use this theorem here by recognizing that $S_1$, $S_2$, $S_3$ and $S_4$ are eigenfunctions of the symmetry operator $\hat{\sigma}$ associated with the plane of symmetry, and that the eigenvalues for $S_1$ and $S_2$ are one and for $S_3$ and $S_4$ are minus one (Exercise 12-4). Now the Hamiltonian commutes with $\hat{\sigma}$, as it does with any symmetry operator (cf. Section 8-7). Therefore, the theorem we have just stated causes many of the elements in the secular determinant to be zero, e.g., $H_{13}^s = \int S_1 \hat{H} S_3 d\tau = 0$ (since $S_1$ and $S_3$ have different eigenvalues). We can, therefore, write Eq. (12-23) as

$$\begin{vmatrix} H_{11}^s - E & H_{12}^s & 0 & 0 \\ H_{21}^s & H_{22}^s - E & 0 & 0 \\ 0 & 0 & H_{33}^s - E & H_{34}^s \\ 0 & 0 & H_{43}^s & H_{44}^s - E \end{vmatrix} = 0 \qquad (12\text{-}24)$$

We can now look in more detail at the remaining elements. If we write the Hamiltonian integrals in Eq. (12-24) in terms of the original basis atomic orbitals and apply the Hückel approximations, we get, for example,

$$H_{11}^s = \int S_1 \hat{H} S_1 d\tau = \tfrac{1}{2} \int (\phi_1 + \phi_4)\hat{H}(\phi_1 + \phi_4)d\tau$$

$$= \tfrac{1}{2}\left[ \int \phi_1 \hat{H}\phi_1 d\tau + \int \phi_1 \hat{H}\phi_4 d\tau + \int \phi_4 \hat{H}\phi_1 d\tau + \int \phi_4 \hat{H}\phi_4 d\tau \right] = \alpha \quad (12\text{-}25)$$

If we do this for all the integrals and use Eq. (12-10), we get

$$\begin{vmatrix} x & 1 & 0 & 0 \\ 1 & x+1 & 0 & 0 \\ 0 & 0 & x & 1 \\ 0 & 0 & 1 & x-1 \end{vmatrix} = 0 \qquad (12\text{-}26)$$

When a determinant can be written like this, with blocks of elements on the diagonals and zeros elsewhere, we can factorize it; we can illustrate what we mean by this by using the case at hand—we get from Eq. (12-26)

$$\begin{vmatrix} x & 1 \\ 1 & x+1 \end{vmatrix}\begin{vmatrix} x & 1 \\ 1 & x-1 \end{vmatrix} = 0 \qquad (12\text{-}27)$$

Therefore,

$$\begin{vmatrix} x & 1 \\ 1 & x+1 \end{vmatrix} = 0 \qquad \begin{vmatrix} x & 1 \\ 1 & x-1 \end{vmatrix} = 0$$

We shall not carry the analysis any further because we have made the point we wished to make: we have reduced the $4 \times 4$ determinantal equation to two $2 \times 2$ equations—this saves us a considerable amount of work. For large molecules the saving can be very significant. For example, the $\pi$-electron calculation for naphthalene would involve solving a $10 \times 10$ determinant if the molecule's symmetry were ignored. By using the symmetry, the $10 \times 10$ determinant can be reduced to more manageable $3 \times 3$ and $2 \times 2$ determinants.[†]

[†]For details of the use of symmetry to solve the HMO problem for naphthalene, see reference B6 in the Bibliography, p. 265.

## SUMMARY

1. In this chapter we very briefly examined a method whereby very approximate, but useful, MO calculations can be made for the $\pi$-electron systems of organic molecules — the method is called *Hückel molecular-orbital (HMO) theory*.

2. We first assumed that the $\pi$-electron system is independent of the $\sigma$-electron system, and applied the LCAO method to the $\pi$-system.

3. HMO theory simplifies the secular determinant resulting from the application of the LCAO method by making some rather drastic approximations in which non-nearest neighbor interactions between atomic orbitals are neglected, as also is overlap of the atomic orbitals.

4. We applied the theory to some very simple $\pi$-electron systems — ethylene, butadiene and benzene, and showed how orbital energies and orbital functions could be obtained for the $\pi$-electrons.

5. We showed, using the results for butadiene as an example, how HMO theory could give information about the $\pi$-electron distribution in a molecule. In this regard, we defined: the *$\pi$-electron charge density* at a particular atom, the *$\pi$-electron bond order* for a particular bond, and *free-valence number*.

6. We showed, using butadiene as an example, how the mathematical computation can be reduced by using the symmetry properties of molecules.

## EXERCISES

**12-1**  Determine the coefficients in Eq. (12-7); i.e., verify that Eqs. (12-13a) and (12-13b) are correct.

**12-2**  If one of the $\pi$-electrons is removed from butadiene to leave the monopositive ion, what is the $\pi$-electron charge density at each carbon atom? What do these results tell you about the distribution of the charge on the ion?

**12-3**  Prove the quantum mechanical theorem stated in the paragraph below Eq. (12-23).

**12-4**  Verify that the functions $S_1$, $S_2$, $S_3$, and $S_4$ in Eq. (12-22) are eigenfunctions of the operator associated with the symmetry operation illustrated in Fig. 12-6, i.e., reflection in the $\sigma$ plane. Also, show that the eigenvalues corresponding to these functions are one for $S_1$ and $S_2$ and minus one for $S_3$ and $S_4$.

**12-5**  Apply the Hückel approximation to all the elements in Eq. (12-24) and thereby show that Eq. (12-26) is correct.

**12-6** Complete the analysis of Section 12-8; i.e., determine expressions for the orbital energies for butadiene in terms of $\alpha$ and $\beta$. Verify that this analysis gives the same results for the orbital energies as those given in Section 12-4 and tabulated in Table 12-1.

# 13

# Bonding in Complexes

## 13-1  INTRODUCTION

Our purpose in this chapter is to show in a rather qualitative way how the ideas we have introduced in previous chapters can be applied to the understanding of bonding in complexes. We shall be especially concerned with applications of the theory and concepts of Chapter 8 (Symmetry in Chemistry).

When we talk about a complex we shall in general mean any arrangement of atoms or ions in which the central atom or ion is surrounded by two or more chemical groups. These latter groups may be ions such as $F^-$ and $CN^-$; or neutral molecules such as $H_2O$ and $NH_3$ — the general name given to these groups is *ligands*. The most important complexes are those of the transition metals — examples are: $[Ni(H_2O)_6]^{2+}$, $[Fe(CN)_6]^{4-}$ and $CrO_4^-$. You will note that in two of our examples the central ion is attached to six ligands; in fact, this situation, in which there is an octahedral arrangement of ligands, is the most common, so we shall use this as our example for the discussions in this chapter. Similar arguments apply to other arrangements of ligands, such as in the tetrahedral case exemplified by $CrO_4^-$.

When dealing with the bonding in complexes, we could apply the regular MO theory; to do this rigorously leads to severe mathematical difficulties. The alternative approaches are to use the more qualitative *crystal-field* and *ligand-field theories*. In crystal-field theory the bonding is considered to be due entirely to the electrostatic attractions between the central cation and the negatively-charged or polar ligands; in ligand-field theory an attempt is made to modify crystal-field theory to allow for overlap of the ligand and central ion orbitals. All the theories are based on the symmetry properties of the complex in question, and it is this aspect of the problem that we shall mainly deal with in this chapter. In short, we shall examine with symmetry arguments what the effect of the surrounding ligands is on the central ion of a complex.

## 13-2  EFFECT OF AN OCTAHEDRAL LIGAND FIELD

The Hamiltonian for a complexed atom is

$$\hat{H} = -\frac{1}{2}\sum_i \nabla_i^2 - \sum_i \frac{Z}{r_i} + \sum_{i<j} \frac{1}{r_{ij}} + \text{spin-orbit coupling term} + V \quad (13\text{-}1)$$

We have previously met all the terms in Eq. (13-1) except the last: the first refers to the kinetic energy of the electrons; the second to the nucleus-electron attractions; the third to the electron-electron repulsions; the fourth to spin-orbit coupling. $V$ is a new term that accounts for the effect of the ligands.

In the Russell-Saunders coupling scheme for a free ion the most important potential energy term is the nucleus-electron attraction term; the next most important is the electron-electron repulsion term; the spin-orbit coupling is less important than these two. The relative effect of $V$ depends on the particular complex considered. If the effect of $V$ is greater than the electron-electron repulsion effect, we have what is known as a *strong-field case*; if it lies between the electron-electron effect and the spin-orbit coupling effect, we have a *weak-field case*. These are the most important cases since they are the operative ones for the first transition series. For the higher transition series $V$ may have less effect than spin-orbit coupling.

We shall specifically consider the first transition series elements, though the general arguments apply to complexes of other elements. The central ion is one of these elements, in which case the orbitals of interest to us are the five $d$ orbitals. In the absence of an external field these orbitals are five-fold degenerate. The question we ask is: What is the effect of an octahedral ligand field, due to six ionic or polar ligands, on these orbitals?

Our method of approach will be to gain a qualitative understanding of the effects of the symmetry properties of complexes by examining some pictorial representations of the situation; then we shall apply the more formal methods of group theory. Our initial approach, as we shall soon see, has some serious shortcomings, but at least it gives an idea of the problems involved.

First, we consider the orientation of the six ligands with respect to the central ion—this is shown diagrammatically in Fig. 13-1. We assume that the ligands are point charges, and that their effect on the $d$ orbitals of the central ion is entirely electrostatic.

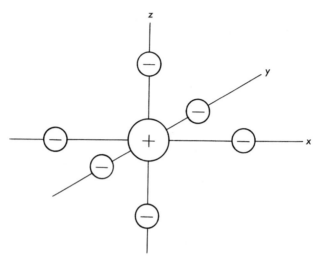

**Fig. 13-1**    Diagrammatic representation of an octahedral ligand field.

Now we recall what the $d$ orbitals look like—these are shown in Fig. 13-2. It should be immediately obvious than the $d_{xy}$, $d_{yz}$, and $d_{xz}$ orbitals are similarly oriented with respect to the ligand field. It should also be obvious that the $d_{z^2}$ and $d_{x^2-y^2}$ orbitals are oriented differently from the other three. Furthermore, we now recall that the $d_{z^2}$ orbital can be regarded as a linear combination of $d_{z^2-x^2}$ and $d_{z^2-y^2}$ orbitals. Since the $d_{x^2-y^2}$, $d_{z^2-x^2}$, and $d_{z^2-y^2}$ orbitals are similarly oriented with respect to the ligand field, it should be intuitively obvious that the $d_{z^2}$ and $d_{x^2-y^2}$ orbitals will have the same energy. We therefore conclude that the $d$ orbitals split in an octahedral field into a three-fold degenerate set and a two-fold degenerate set.

In the point-charge model we would expect the $d_{xy}$, $d_{yz}$ and $d_{xz}$ orbitals to have lower energy than the other two—this is because the orientations of the former orbitals allow electrons in them to come less under the influence of the negative charges on the ligands. (To see this you should mentally combine Figs. 13-1 and 13-2.) We show the splitting of the $d$ orbitals in Fig. 13-3. (We explain the symbols beside the orbitals in Fig. 13-3 when we apply group theory to the problem.)

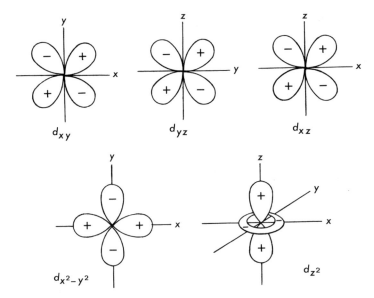

**Fig. 13-2**   Three-dimensional representation of the five $d$ orbitals.

The argument we have just given does give the right qualitative results. But it has some serious deficiencies. For example, we must object to treating the ligands as point charges. In fact, the ligand orbitals overlap the $d$ orbitals of the central ion — this results in bonds with covalent character. Even in complexes where we would expect the bonding to be very

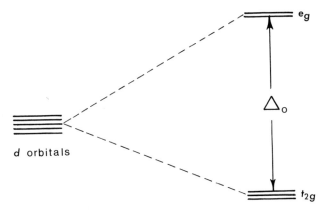

**Fig. 13-3**   Splitting of the $d$ orbitals in an octahedral field. The labels $e_g$ and $t_{2g}$ are the group theory symbols resulting from the transformation properties of the two sets of orbitals.

ionic (e.g., $CoF_6^{3-}$) a significant amount of covalence occurs. Thus, we expect that the occupied bonding orbitals will not be pure $d$ orbitals, and that the delocalization of the electrons into the ligand orbitals will occur. We shall not pursue this matter further. For a simple summary of the qualitative implications of orbital overlap you can consult reference F8 in the Bibliography. Calculations that take into account the effects of covalent bonding show that our original result regarding the splitting of the degeneracy of the $d$ orbitals is right.

## 13-3  USE OF GROUP THEORY

In this section we show how the ideas we discussed in Chapter 8 can be used to examine in a formal way the effects of the symmetry properties of complexes. We continue to use the important case of an octahedral ligand arrangement and its effect on the $d$ orbitals of a central ion; we can therefore continue to refer to the diagrams we have already used.

In the last section we examined the splitting of the degeneracy of the $d$ orbitals. What we really wish to do is examine the effect on the total electronic states of the central ion. For example, an ion with $d^2$ configuration can give, using the Russell-Saunders coupling method, the terms $^1S$, $^1D$, $^1G$, $^3P$, $^3F$; by Hund's rules, the ground-state term is $^3F$. An ion with $d^1$ configuration gives just a $^2D$ term, we shall use the latter case as an illustration.

A $D$ state of an ion has the same symmetry properties as the $d$ orbitals we discussed in the last section. We can see this by examining the form of the $d$ orbital functions as given by Eq. (6-104) with $l = 2$; i.e.,

$$\psi_d = R(r)Y_{2,m}(\theta, \phi) \qquad (13\text{-}2)$$

We shall see in a moment that we need only examine the $\phi$ dependence of this function; this takes the form exp $(im\phi)$ (cf. Eq. (6-62)). The five-fold degeneracy of the $d$ orbitals arises because $m$ can take the values $2, 1, 0, -1, -2$. Now the function for a $D$ atomic term has the same $\phi$ dependence, i.e., is given by exp $(iM_L\phi)$; again there is five-fold degeneracy because $M_L$ can take five values. We conclude from this argument that any symmetry effects on the $d$ orbitals due to the ligand environment will be the same as on the five components of a $D$ electronic state. In fact, we can extend the argument to other terms and orbitals — $s$ orbitals and $S$ states are related; $p$ orbitals and $P$ states are related; and so on. We shall here develop the theory in terms of the orbital functions.

We now use the functions we have been discussing as a basis for a representation of the point group associated with the octahedral environment created by the ligands. The representation will be reducible; by breaking it down into its component irreducible representations, we can get some useful information about the splitting of the five-fold degeneracy when the ion is influenced by the ligands.

The octahedral point group is $O_h$. However, we can work with the simpler $O$ group. The symmetry operations of the $O$ group are $E$, $6C_4$, $8C_3$, $3C_2$, and $6C_2'$. To get the $O_h$ group from this, all we have to do is add the inversion operator and its combinations with the rotations. Here we are dealing with $d$ orbitals, and the inversion symmetry of these is obvious (cf. Fig. 13-2) – only $g$ terms can arise.

Our objective is to find the characters of the transformation matrices for the operations of the $O$ group on the $d$ orbital functions. We can then use these characters, the character table for the $O$ group and Eq. (8-31) to determine the irreducible representations into which the reducible representation breaks down. This will show us that the five orbitals (or components of terms if this is what we are considering) can be split into two sets, each set acting differently in the octahedral field.

For convenience we define the $z$-axis as the rotation axis for the rotation operation we happen to be considering. By doing this, we see that the rotations cannot affect the parts of the functions Eq. (13-2) that depend on $\theta$ because $\theta$ is measured from the $z$-axis. $R(r)$ will also not be affected by the rotations, so we need only consider the parts of the functions with $\phi$ dependence, i.e., $\exp(im\phi)$. If we perform a rotation by $\alpha$, this changes to $\exp(im(\phi + \alpha))$. We can show the effect on all five functions by the equation

$$\hat{C}(\alpha) \begin{bmatrix} e^{2i\phi} \\ e^{i\phi} \\ e^0 \\ e^{-i\phi} \\ e^{-2i\phi} \end{bmatrix} = \begin{bmatrix} e^{2i\alpha} & 0 & 0 & 0 & 0 \\ 0 & e^{i\alpha} & 0 & 0 & 0 \\ 0 & 0 & e^0 & 0 & 0 \\ 0 & 0 & 0 & e^{-i\alpha} & 0 \\ 0 & 0 & 0 & 0 & e^{-2i\alpha} \end{bmatrix} \begin{bmatrix} e^{2i\phi} \\ e^{i\phi} \\ e^0 \\ e^{-i\phi} \\ e^{-2i\phi} \end{bmatrix}$$

$$= \begin{bmatrix} e^{2i(\phi + \alpha)} \\ e^{i(\phi + \alpha)} \\ e^0 \\ e^{-i(\phi + \alpha)} \\ e^{-2i(\phi + \alpha)} \end{bmatrix} \tag{13-3}$$

The characters of the transformation matrices are therefore given by

$$\chi(\alpha) = e^{2i\alpha} + e^{i\alpha} + e^0 + e^{-i\alpha} + e^{-2i\alpha}$$

Remembering that $\exp(ix) = \cos(x) + i \sin(x)$, we can write this

$$\chi(\alpha) = 2 \cos 2\alpha + 2 \cos \alpha + 1 \qquad (13\text{-}4)$$

We therefore get the characters of the operations of the $O$ group by substituting $\alpha = 0$ for $E$, $\alpha = \pi/2$ for $C_4$, and so on; doing this, we get

$$
\begin{array}{c|ccccc}
 & E & C_4 & C_3 & C_2 & C_2' \\
\hline
\chi(R) & 5 & -1 & -1 & 1 & 1
\end{array}
\qquad (13\text{-}5)
$$

**Table 13-1**   Character table for the $O$ group.

| $O$ | $E$ | $6C_4$ | $8C_3$ | $3C_2$ | $6C_2'$ |
|-----|-----|--------|--------|--------|---------|
| $A_1$ | 1 | 1 | 1 | 1 | 1 |
| $A_2$ | 1 | −1 | 1 | 1 | −1 |
| $E$ | 2 | 0 | −1 | 2 | 0 |
| $T_1$ | 3 | 1 | 0 | −1 | −1 |
| $T_2$ | 3 | −1 | 0 | −1 | 1 |

The representation in Eq. (13-5) is reducible; we can find the irreducible representations it contains by using Eq. (8-31) and the character table for the $O$ group (Table 13-1). Equation (8-31) is

$$a_j = \frac{1}{h} \sum_R \chi(R)\chi_j(R)$$

Considering just one example, we shall calculate the number of times the $T_2$ representation occurs:

$$a_{T_2} = \frac{1}{24}[(1 \times 5 \times 3) + (6 \times -1 \times -1) + (8 \times -1 \times 0)$$

$$+ (3 \times 1 \times -1) + (6 \times 1 \times 1)] = 1$$

You can complete the rest of the analysis as an exercise (Exercise 13-4). The result is

$$\Gamma_d = e_y + t_{2y} \qquad (13\text{-}6)$$

Here we have used the small letters $e$ and $t$ rather than $E$ and $T$ — this is conventional notation for referring to one-electron orbitals. Also, you will note that the analysis did not give the $g$ subscripts; we have already mentioned that these must be $g$ because of the inversion symmetry of the $d$ orbitals.

What does the result expressed in Eq. (13-6) mean? It means that in an octahedral ligand field the central ion's $d$ orbitals are split into two sets.

One set, which is doubly degenerate, transforms according to the $e_g$ representation; the other, which is triply degenerate, transforms according to the $t_{2g}$ representation.

If in this analysis we had used the components of $D$ states instead of the $d$ orbitals, we would have concluded that these also split into two sets, transforming according to $E_g$ and $T_{2g}$ representations. We can use similar arguments to determine the effects of ligand fields on other orbitals and other atomic states. If you are interested in the results for these other cases, you can consult, for example, reference F 2 in the Bibliography.

## 13-4  CORRELATION DIAGRAMS

The question we ask in this section is: What will be the distribution of the electrons in the $d$ orbitals of the complex? Now that the degeneracy no longer exists, the probability of the electrons' occupying them will not be equal. Two factors are operative: the Pauli exclusion principle and electron-electron repulsions. The first of these factors operates in the same way as for free atoms and ions, so we shall say nothing further about it. The effect of the second factor will depend on the strength of the ligand field. It is our intention to construct a diagram that shows the varying effect of the ligand field on the states of the complexed atom or ion. We choose as an example the $d^2$ case.

If the ligand field is very strong, the value of $\Delta_0$ in Fig. 13-3 will be very large. The possible configurations for the electrons will be, in order of increasing energy, $t_{2g}^2$, $t_{2g}e_g$, $e_g^2$. If the field is less than infinitely strong, the electrons will feel each other's influence—the configurations we have noted above will then give rise to a set of terms. To determine these we take the direct products (Section 8-8) and use Eq. (8-31) to analyze the symmetry properties of the terms. We show the results on the right side of Fig. 13-4—the connecting lines show the terms given by the infinite-field configurations when the field is relaxed. In this diagram we have shown the multiplicities of the terms—we shall not here go into the details of working these out.

We now combine these results with those for a free ion and the terms associated with a weak ligand field. We have already noted that for a free ion with $d^2$ configuration the possible terms are, in increasing order of energy, $^3F$, $^1D$, $^3P$, $^1G$, $^1S$. We have shown that a $D$ term in a weak field gives $E_g$ and $T_{2g}$ terms. We could do the same for the other states. The results, again including the multiplicities, are shown on the left side of Fig. 13-4.

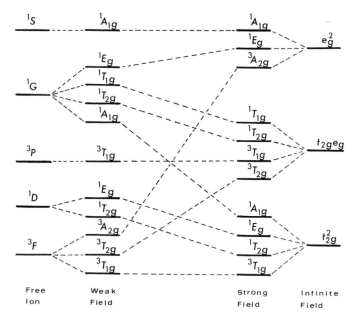

**Fig. 13-4** Correlation diagram for a $d^2$ configuration in an octahedral field.

Figure 13-4 is the diagram we wished to construct — it is called a *correlation diagram*. To get it we have made use of two principles that we shall state without proof:

1. There must be a one-to-one correspondence between states on the two sides of the diagram.

2. The *non-crossing rule*, which says that states of the same multiplicity and symmetry cannot cross.

At the far left of the diagram are the states of the free ion. To the right of these are the states into which they split when the ion is in an octahedral field. On the far right are the states for an infinitely strong ligand field, and to the left of these are the states into which these split if the field is weakened somewhat but remains very strong. Using the two principles stated above, we can draw the correlation lines only as shown; i.e., as the interaction with the ligand field increases the states on the left must go into the states on the right, and the non-crossing rule tells us the exact way in which they correlate.

## 13-5  CONCLUDING REMARKS

In this chapter we have restricted our discussions to the single case of an octahedral complex in which the $d$ orbitals of the central ion are the important bonding orbitals. The same sort of arguments can be applied to other types of complexes—for example, square planar and tetrahedral complexes. The important point to make is that, though rather qualitative, crystal-field and ligand-field theories are very useful in that they correlate a tremendous quantity of experimental results. Examples of these experimental results are: spectroscopic measurements, magnetic properties of complexes and thermodynamic results such as measurements of bond energies—many of the more important of these observations can be explained and correlated using the results of crystal-field and ligand-field theories.

## SUMMARY

1. In this chapter we examined the bonding in complexes in terms of symmetry arguments.

2. We chose as an example the complexes of the first transition series in which the central ions were in octahedral ligand fields. In this case it is the $d$ orbitals of the central ion that are important in the bonding theory.

3. We saw that in the octahedral field the five-fold degeneracy of the $d$ orbitals is split. Very qualitative, pictorial arguments and the more formal arguments of group theory show the splitting to be into two sets, one doubly degenerate and the other triply degenerate. Group theory showed the former to be of $e_g$ symmetry and the latter to be of $t_{2g}$ symmetry.

4. Atomic states are split similarly to the analogous orbitals; e.g., $D$ states and $d$ orbitals are affected in similar ways by the ligand environment.

5. Using the symmetry property results, we saw how we could construct a *correlation diagram* which gives the effect on the electronic states of the central ion when the ligand field increases from zero, through weak and strong, to infinitely strong.

## EXERCISES

**13-1**  By considering the geometry of a tetrahedral complex, as shown in Fig. 13-5, and by using the same type of arguments as were used in Section 13-2 to give Fig. 13-3, draw an energy level diagram to show how the $d$ orbitals are split in a tetrahedral field.

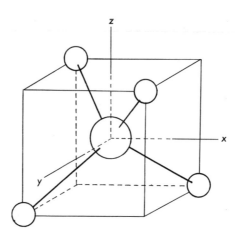

**Fig. 13-5**

**13-2** Work out the details of adding the $i$ operation to the operations of the $O$ group; i.e., determine the additional symmetry operations that occur in the $O_h$ group.

**13-3** Show that an $F$ state arising from, for example, a $d^2$ configuration splits in an octahedral field to give $T_{1g}$, $T_{2g}$ and $A_{2g}$ components. (Hint: Use the general result that the character for a rotation by $\alpha$ is given by

$$\chi(\alpha) = \frac{\sin (L + 1/2)\alpha}{\sin \alpha/2}$$

for $\alpha \neq 0$.)

**13-4** Using Table 13-1, Eq. (8-31) and the reducible representation of Eq. (13-5), complete the analysis that resulted in Eq. (13-6).

# 14

# Spectroscopy

## 14-1  INTRODUCTION

It is in the field of spectroscopy that much of the experimental evidence for the validity of quantum chemical theory has been obtained; spectroscopic results have also given much incentive for the formulation of many of the fundamental ideas of quantum theory. We would be remiss, therefore, if we did not at least briefly mention this topic. However, in a book of this type it is quite impossible for us to give anything like an adequate coverage of even the very fundamental theory. What we shall do then is examine the most basic theory of spectroscopy and leave it at that. For further treatment of the subject you will have to consult one of the books on spectroscopy listed at the end of this book.

So far in this book we have been concerned with stationary states of systems; our theoretical treatment has been in terms of the time-independent Schroedinger equation. In spectroscopy we deal with transitions between stationary states, and to examine these we need the more general time-dependent Schroedinger equation.

We shall first derive the equation that gives the rate of change of a chemical system from one quantum mechanical state to another when it is influenced by a perturbing effect. Then we shall apply this equation to the understanding of spectroscopic phenomena, in which case the perturbation is brought about by electromagnetic radiation.

## 14-2  TIME-DEPENDENT PERTURBATIONS

Let us first recall the time-dependent Schroedinger equation, Eq. (4-3); it is

$$\hat{H}\Psi(q,t) = -\frac{\hbar}{i}\frac{\partial}{\partial t}\Psi(q,t) \tag{14-1}$$

$\Psi(q,t)$, remember, is a function that is time-dependent.

We shall assume for the sake of simplicity that the system we are dealing with has just two stationary states, one state that we shall label $m$ and a state with higher energy that we shall label $n$. The time-dependent functions $\Psi_m$ and $\Psi_n$ are then given by Eq. (4-78); i.e.,

$$\Psi_m = \psi_m(q)\exp\left(-\frac{iE_m t}{\hbar}\right) \tag{14-2a}$$

$$\Psi_n = \psi_n(q)\exp\left(-\frac{iE_n t}{\hbar}\right) \tag{14-2b}$$

where $\psi_m$ and $\psi_n$ are time-independent functions; $E_m$ and $E_n$ are the energies associated with these functions.

Since we are considering a system with only two stationary states, the general solution to Eq. (14-1) can be expressed as

$$\Psi = a_m\Psi_m + a_n\Psi_n \tag{14-3}$$

where the coefficients $a_m$ and $a_n$ are functions of time.

We now write the Hamiltonian

$$\hat{H} = \hat{H}^0 + \hat{H}' \tag{14-4}$$

$\hat{H}^0$ is the Hamiltonian for an unperturbed system; $\hat{H}'$ is the time-dependent perturbation term. We assume that the effect of $\hat{H}'$ is small compared to that of $\hat{H}^0$. Substituting Eqs. (14-3) and (14-4) into Eq. (14-1), we get

$$(\hat{H}^0 + \hat{H}')(a_m\Psi_m + a_n\Psi_n) = -\frac{\hbar}{i}\frac{\partial}{\partial t}(a_m\Psi_m + a_n\Psi_n) \tag{14-5}$$

i.e.,

$$a_m\hat{H}^0\Psi_m + a_n\hat{H}^0\Psi_n + a_m\hat{H}'\Psi_m + a_n\hat{H}'\Psi_n$$

$$= -\frac{\hbar}{i}\frac{da_m}{dt}\Psi_m - \frac{\hbar}{i}a_m\frac{\partial\Psi_m}{\partial t} - \frac{\hbar}{i}\frac{da_n}{dt}\Psi_n - \frac{\hbar}{i}a_n\frac{\partial\Psi_n}{\partial t} \tag{14-6}$$

Now $\Psi_m$ and $\Psi_n$ are solutions of the Schroedinger equation for the unperturbed system; i.e., they satisfy the equation

$$\hat{H}^0\Psi = -\frac{\hbar}{i}\frac{\partial\Psi}{\partial t} \tag{14-7}$$

Therefore, the first two terms on the left of Eq. (14-6) cancel the second and last terms on the right. If we multiply through by $\Psi_n^*$ and integrate over all space, we get

$$a_m\int \Psi_n^*\hat{H}'\Psi_m d\tau + a_n\int \Psi_n^*\hat{H}'\Psi_n d\tau$$
$$= -\frac{\hbar}{i}\frac{da_m}{dt}\int \Psi_n^*\Psi_m d\tau - \frac{\hbar}{i}\frac{da_n}{dt}\int \Psi_n^*\Psi_n d\tau \tag{14-8}$$

The first integral on the right is zero; the second integral is one – this is because of the orthonormality of the wavefunctions. If we substitute Eqs. (14-2a) and (14-2b) into Eq. (14-8) and rearrange, we get

$$\frac{da_n}{dt} = -\frac{i}{\hbar}a_m\exp\left(-\frac{i(E_m-E_n)t}{\hbar}\right)H'_{nm} - \frac{i}{\hbar}a_nH'_{nn} \tag{14-9}$$

where

$$H'_{nm} = \int \psi_n^*\hat{H}'\psi_m d\tau$$

Initially, $a_n$ is zero, so the last term on the right of Eq. (14-9) would be zero. We could show, though we shall not do so, that this term is not very significant even at longer times; therefore, since $a_m$ remains approximately one, we can write Eq. (14-9) as

$$\frac{da_n}{dt} = -\frac{i}{\hbar}\exp\left(\frac{i(E_n-E_m)t}{\hbar}\right)H'_{nm} \tag{14-10}$$

This is the equation we set out to get; since $a_n$ is the coefficient for the state of higher energy, this equation gives a measure of the rate of transition to this state when the system is influenced by the perturbing effect.

We could have carried the argument through for the more general case where more than two stationary states exist for the system. We would then have had to write instead of Eq. (14-3):

$$\Psi = \sum_r a_r\Psi_r$$

With this function we would have obtained instead of Eq. (14-10):

$$\frac{da_n}{dt} = -\frac{i}{\hbar}\sum_r a_r\exp\left(\frac{i(E_n-E_r)t}{\hbar}\right)H'_{nr} \tag{14-11}$$

(The derivation of Eq. (14-11) is left for you to do as an exercise.) We shall not use this general equation in what follows, but shall continue to use the restricted Eq. (14-10).

## 14-3   RADIATION THEORY

In this section we shall consider the case where the perturbation we discussed in the last section is caused by interaction of the chemical system with electromagnetic radiation — we shall derive the equation giving the rate of change of a chemical system from one state to another when it is acted on by an electromagnetic radiation. First, we consider the interaction with the $x$ component of the radiation; then we generalize the result.

Electromagnetic radiation, as its name implies, has oscillating electric and magnetic fields associated with it. In order to illustrate this, we can consider a plane-polarized wave as shown in Fig. 14-1. In this case the

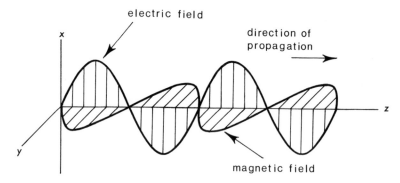

**Fig. 14-1**  Oscillating electric and magnetic field vectors for a plane polarized electromagnetic wave. The electric and magnetic fields are perpendicular to each other and to the direction of propagation of the wave.

electric and magnetic field vectors are each oriented in one plane, as shown — the electric and magnetic fields are perpendicular to each other and to the direction of the propagation of the wave (the $z$ direction in this case).

Here we shall consider that electromagnetic radiation (which in general will, of course, be unpolarized) interacts with an atom or molecule. For simplicity we shall consider only the interaction of the electric field since this is the important interaction in most kinds of spectroscopy. (However,

This result refers to a single absorption frequency. In practice, absorption will take place over a range of frequencies. However, unless the condition Eq. (14-18) is approached, absorption will not be significant; i.e., the range of frequencies absorbed will be small enough for us to consider $E_x^0$ constant. We can therefore integrate over all frequencies and get the result

$$a_n^* a_n = |\mu_{xmn}|^2 \frac{(E_x^0)^2 t}{\hbar^2} \tag{14-23}$$

To get this result we have substituted $x = (E_n - E_m - h\nu)t/2\hbar$ and used the integral

$$\int_{-\infty}^{+\infty} \frac{\sin^2 x}{x^2} dx = \pi$$

If we differentiate Eq. (14-23) with respect to time, we get the rate of transition from the lower state $m$ to the upper state $n$:

$$\frac{d(a_n^* a_n)}{dt} = |\mu_{xmn}|^2 \frac{(E_x^0)^2}{\hbar^2} \tag{14-24}$$

Here we are expressing the rate of transition in terms of the electric field component $E_x^0$. It is more usual to express it in terms of radiation energy density $\rho$. If we do this, we get

$$\frac{d(a_n^* a_n)}{dt} = \frac{2\pi}{3\hbar^2} |\mu_{xmn}|^2 \rho \tag{14-25}$$

Remember that we have considered only the $x$ component of the radiation. If we assume the radiation to be isotropic, we can include the other components as follows:

$$\frac{d(a_n^* a_n)}{dt} = \frac{2\pi}{3\hbar^2} |(\mu_{xmn})^2 + (\mu_{ymn})^2 + (\mu_{zmn})^2| \rho$$

$$= \frac{2\pi}{3\hbar^2} |\mu_{mn}|^2 \rho \tag{14-26}$$

$|\mu_{mn}|$ is known as the *transition moment*. Finally, we write Eq. (14-26) using *Einstein's coefficient for induced absorption* $B_{mn}$ given by

$$B_{mn} = \frac{2\pi}{3\hbar^2} |\mu_{mn}|^2 \tag{14-27}$$

This gives

$$\frac{d(a_n^* a_n)}{dt} = B_{mn}\rho \tag{14-28}$$

There are some points we can make regarding the theory we have just examined:

1. Equation (14-28) shows us that the rate of transition between the two states depends on the transition moment and on the energy density of the radiation.

2. Significant absorption of radiation occurs only if the frequency is such that the condition expressed in Eq. (14-18) is approached, because it is only under this condition that $a_n$ becomes large. We should recall that Eq. (14-18) expresses the Bohr condition for absorption of radiation.

3. We can easily relate our theory to experimental measurements. According to *Beer's law*, the change in intensity of radiation of frequency $\nu$ when it is absorbed by a sample of thickness $dl$ and molar concentration $C$ (cf. Fig. 14-2) is given by

$$-dI = \epsilon(\nu)ICdl \tag{14-29}$$

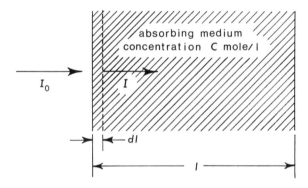

**Fig. 14-2**   Absorption of radiation by an absorbing medium. The initial intensity is $I_0$; $I$ is the intensity after some absorption has occurred.

Here $I$ is the intensity of radiation (measured as energy per unit area); $\epsilon(\nu)$, which is dependent on the frequency, is the *molar extinction coefficient* or *absorption coefficient*. If we integrate Beer's law over the total thickness of the absorbing sample $l$, we get

$$\epsilon(\nu) = \frac{1}{Cl}\ln\frac{I_0}{I} \tag{14-30}$$

$I_0$ is the initial intensity of the radiation. In practice, absorption usually occurs over a band of frequencies. We therefore define the *integrated absorption coefficient A*:

$$A = \int_{\nu_a}^{\nu_b} \epsilon(\nu)d\nu \tag{14-31}$$

This integration is carried out over the whole absorption band where the lowest and highest frequencies in the band are $\nu_a$ and $\nu_b$.

The integrated absorption coefficient is an experimentally measurable quantity, so we would like to relate it to our theoretically derived quantity $B_{mn}$ as given by Eq. (14-27). Now the amount of radiant energy in the frequency range $\nu_a$ to $\nu_b$ that is absorbed by the sample per unit volume per unit time is given by

$$ICA = IC \int_{\nu_a}^{\nu_b} \epsilon(\nu)d\nu \tag{14-32}$$

Every time a transition takes place an amount of energy $h\nu_{mn}$ is absorbed from the radiation (where $\nu_{mn}$ is the frequency of the absorbed radiation). Therefore, according to Eq. (14-28), the energy absorbed per unit volume per unit time for this particular frequency is given by

$$B_{mn}\rho h\nu_{mn}N' \tag{14-33}$$

where $N'$ is the number of molecules per cubic centimeter of sample:

$$N' = \frac{N_0 C}{1000} \qquad (N_0 = \text{Avogadro's Number})$$

If we assume the energy is absorbed over a narrow range of frequencies, we can equate Eqs. (14-32) and (14-33) to get

$$A = \frac{B_{mn}\rho h\nu_{mn}N_0}{1000 I} \tag{14-34}$$

Finally, we can recognize that $I$ is the energy flowing through a cross-sectional area of 1 cm² in 1 sec. $I$ is therefore related to the energy density $\rho$ by

$$I = c\rho$$

where $c$ is the velocity of propagation of the radiation. Using Eq. (14-27), Eq. (14-34) therefore becomes

$$A = \frac{8\pi^3 N_0}{3000 \, hc} \nu_{mn}|\mu_{mn}|^2 \tag{14-35}$$

This is the expression we set out to obtain because it gives the relationship between the experimentally measurable integrated absorption coefficient and the theoretically defined transition moment.

## 14-4 EXPERIMENTAL SPECTROSCOPY

In the remaining sections of this chapter we shall consider some of the simpler concepts involved in the interpretation of experimental spectroscopic results. The discussion is certainly not meant to be a coverage of all the basic questions that we have to consider when interpreting spectroscopic measurements; rather, it will serve to give some indication as to how quantum mechanical ideas, as discussed in the early chapters of this book and in this chapter, are used to explain experimental results in the very important field of spectroscopy. We shall limit our discussion to diatomic molecules since these systems show sufficiently complex spectra to illustrate many of the basic ideas that have to be understood in order to examine molecular spectra. In general, the mathematics needed to understand the spectra of more complex molecules is much more complicated, so we shall not consider polyatomic molecules here.

We recall from our cursory examination of the historical background to quantum theory (cf. Chapter 2) that it was in the field of atomic spectroscopy that many of the early quantum mechanical ideas found experimental justification. In general, atomic spectra consist of series of lines resulting from electronic transitions between the allowed electronic states of the atom in question. In Fig. 2-1 we illustrated some possible electronic transitions for the simplest atom, hydrogen. We could draw similar diagrams for other atoms, thereby explaining the line spectra that are characteristic of atomic spectroscopy.

Molecules, unlike atoms, give spectra that are complicated by the possibility of a great variety of transitions associated with rotational and vibrational energies. The transitions between rotational and vibrational states of molecules lead to much fine structure in the spectra, the lines often being so closely spaced as to appear as continuous bands.

Here we shall consider three main types of molecular spectra — pure rotational spectra, vibration-rotation spectra and electronic spectra.

## 14-5 PURE ROTATIONAL SPECTRA

As the name implies, pure rotational spectra result from transitions between the rotational states of molecules. To obtain pure rotational spectra we use energy that is insufficient to induce transitions between vibrational or electronic states, i.e., we are concerned with rotational transitions within a single vibrational state associated with some particular electronic state of a molecule. The separation of rotational states is such that the quanta of radiative energy needed to bring about transitions lie in the microwave or far infra-red regions of the electromagnetic spectrum.

As mentioned in the last section, we shall confine our discussions to the simple case of diatomic molecules. However, the theory we shall develop is also applicable to other linear molecules. As a first approximation, we can consider a diatomic molecule to be a rigid rotator. In Sections 6-2 and 6-3 we examined the rigid rotator problem in some detail; we determined expressions for the allowed energies (cf. Eq. (6-46)) and the corresponding wavefunctions (cf. Eq. (6-62)). We can now use these results to investigate the nature of rotational transitions.

First we examine the transition moment associated with rotational transitions of diatomic molecules. The transition moment is given by (cf. Eqs. (14-16) and (14-26)):

$$|\mu_{mn}| = \int \psi_n^* \mu \psi_m d\tau \tag{14-36}$$

Here $\psi_n$ and $\psi_m$ are solutions of the rigid rotator problem. According to Eq. (6-62), if we use polar coordinates to define the system, $\psi_n$ and $\psi_m$ are functions of $\theta$ and $\phi$. To examine the transition moment we need to express the dipole moment $\mu$ in polar coordinates. Let us therefore represent the dipole moment as in Fig. 14-3. We can then write for its components:

$$\mu_x = \mu \sin \theta \cos \phi$$
$$\mu_y = \mu \sin \theta \sin \phi \tag{14-37}$$
$$\mu_z = \mu \cos \theta$$

This in turn allows us to write expressions for the transition moment components — they are:

$$|\mu_{xmn}| = \mu \int \psi_n^* \sin \theta \cos \phi \psi_m \sin \theta \, d\theta d\phi$$

$$|\mu_{ymn}| = \mu \int \psi_n^* \sin \theta \sin \phi \psi_m \sin \theta \, d\theta d\phi \tag{14-38}$$

$$|\mu_{zmn}| = \mu \int \psi_n^* \cos \theta \psi_m \sin \theta \, d\theta d\phi$$

From this we can see that the transition moment is zero unless the permanent dipole moment of the molecule has a non-zero value. Therefore, molecules such as HCl give pure rotational spectra whereas molecules such as $N_2$ do not.

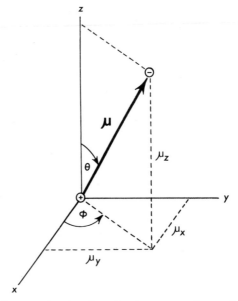

**Fig. 14-3** The dipole moment vector of a molecule. $\mu_x$, $\mu_y$, and $\mu_z$ are the values of the components along the coordinate axes.

The next question to consider is: What are the energies for rotational transitions of diatomic molecules considered as rigid rotators? According to Eq. (6-46), the energies $E_J$ of the allowed rotational states are given by:

$$E_J = J(J+1)\frac{\hbar^2}{2I} \qquad J = 0, 1, 2, 3, \ldots \qquad (14\text{-}39)$$

(Here we have substituted $J$ instead of the quantum number $l$, since this is the conventional symbol to use for molecular rotational states.) A transition between two rotational states, an upper state labeled by a single prime and a lower state labeled by a double prime, therefore involves a change in energy given by:

$$E_{J'} - E_{J''} = \frac{\hbar^2}{2I}[J'(J'+1) - J''(J''+1)] \qquad (14\text{-}40)$$

Although we have written Eq. (14-40) in a way that implies that any integral values can be substituted for $J'$ and $J''$, this is not so. We shall not prove it, but in fact transitions occur only if

$$\Delta J = \pm 1 \qquad (14\text{-}41)$$

This is an example of a *selection rule*. In absorption spectroscopy, the effective part of the rule is $\Delta J = +1$. If we now apply this selection rule to Eq. (14-40), the expression for the energy of a rotational transition becomes:

$$E_{J'} - E_{J''} = \frac{\hbar^2}{2I}[(J''+1)(J''+2) - J''(J''+1)]$$

$$= \frac{\hbar^2}{2I} \cdot 2(J''+1) \qquad (14\text{-}42)$$

We can equate this energy difference to the energy of the quantum of radiation needed to bring about the transition, i.e., $h\nu$ (where $\nu$ is the frequency of the radiation). Therefore, Eq. (14-42) gives us

$$\nu = \frac{h}{8\pi^2 I} \cdot 2(J''+1) \qquad (14\text{-}43)$$

In spectroscopic work it is usual to use wave-number units. Expressing Eq. (14-43) in wave-number units, we get

$$\bar{\nu}\,(\text{cm}^{-1}) = \frac{\nu}{c} = \frac{h}{8\pi^2 Ic} \cdot 2(J''+1)$$

$$= 2\bar{B}(J''+1) \qquad (14\text{-}44)$$

Here we have used the conventional symbol $\bar{B}$ for the constant given by

$$\bar{B} = \frac{h}{8\pi^2 Ic} \qquad (14\text{-}45)$$

$\bar{B}$ is known as the *rotational constant*.

By substituting the values $J = 0, 1, 2, 3, \ldots$ in Eq. (14-44), we can draw an energy level diagram for the rigid rotator approximation to the diatomic molecule – the diagram we get is shown in Fig. 14-4. It is obvious from this figure that the spacing between the lines in the rotational spectrum is constant and is equal to $2\bar{B}$. By using Eq. (14-45) we can therefore calculate a value for the moment of inertia of the molecule (and, consequently, a value for the internuclear distance).

Before we leave this discussion of pure rotational spectra of diatomic molecules we should briefly examine the approximation we made, i.e., that a diatomic molecule can be regarded as a rigid rotator. In fact, the molecule is a non-rigid rotator since the greater the rotational energy, the greater the effect on the bond of the centrifugal forces. These forces lead to *centrifugal stretching* with the consequent change in the moment

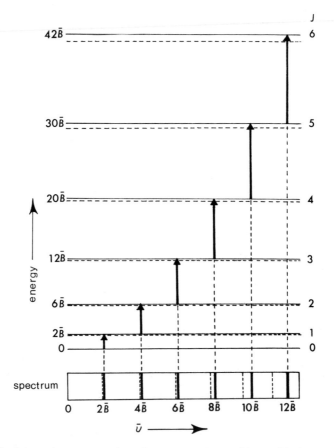

**Fig. 14-4**  Schematic representation of spectroscopic transitions with the resulting spectrum for the rigid rotator. The dashed energy levels and dashed lines in the spectrum indicate the modifications needed if the rotator is considered to be non-rigid; for clarity, the transitions for the non-rigid case are not shown.

of inertia of the molecule. We shall not examine the details of taking centrifugal stretching into account, but shall just note that the expression for the rotational energy levels has to be modified as follows:

$$\bar{E}_J = \bar{B}J(J+1) - \bar{D}J^2(J+1)^2 \qquad (14\text{-}46)$$

Here $\bar{D}$ is the *centrifugal stretching constant*; it is dependent on the force constant of the molecule. We might have expected that the force constant would be involved here since this is a measure of the resistance of the molecule to centrifugal stretching. Usually, $\bar{D}$ is much smaller than $\bar{B}$, so that the effects of centrifugal stretching become important only for large values of $J$.

If we allow for centrifugal stretching, the lines in the rotational spectrum are no longer evenly spaced; in fact, the spacings become progressively smaller as $J$ increases. Figure 14-4 gives an indication of the modifications to the energy levels and to the spectral lines that result from allowance for the fact that the molecule is a non-rigid rotator.

## 14-6  VIBRATION-ROTATION SPECTRA

The quanta of energy needed to bring about transitions between vibrational states of molecules are, in general, larger than those required for rotational transitions. The spectra that result from vibrational transitions therefore occur in the near infra-red region of the electromagnetic spectrum. When considering vibrational transitions, we have to take into account the fact that the vibrational states have rotational states associated with them; therefore, when a molecule interacts with radiation of sufficient energy to bring about vibrational transitions, there are many possibilities for rotational changes to take place also.

Again we consider diatomic molecules. In order to consider the rotational changes, we continue to use the rigid rotator approximation. To consider the vibrational changes we also make an approximation by assuming that the molecule is a harmonic oscillator. We examined the harmonic oscillator in Section 6-4; there we determined expressions for the allowed energies (cf. Eq. (6-92)) and for the wavefunctions (cf. Eq. (6-88)).

As in the case of pure rotational spectra, we shall first examine the transition moment to determine whether there are any conditions that the dipole moment of a molecule has to meet for vibrational transitions to be possible. Since the lowest vibrational state, which we shall label $\psi_0$, is the only appreciably populated state at room temperature, we shall consider a transition from this state to another state $\psi_n$. The transition moment is given by:

$$|\mu_{0n}| = \int \psi_n^* \mu \psi_0 d\tau \tag{14-47}$$

In order to examine this transition moment, we need some knowledge about the dipole moment. The dipole moment is a function of the internuclear distance $R$, but in general we do not know the exact relationship. We therefore consider only small displacements from the equilibrium internuclear distance $R_e$ and express the dipole moment as a series in terms of the displacement coordinate $q$ ($q = R - R_e$):

$$\mu = \mu_0 + \left(\frac{d\mu}{dq}\right)_{q=0} q + \cdots \tag{14-48}$$

Here $\mu_0$ is the permanent dipole moment of the molecule. Substituting Eq. (14-48) in Eq. (14-47) and neglecting all terms other than the first two, we get

$$|\mu_{0n}| = \int \psi_n^* \mu_0 \psi_0 d\tau + \int \psi_n^* \left(\frac{d\mu}{dq}\right)_{q=0} q \psi_0 d\tau \tag{14-49}$$

Now $\psi_0$ and $\psi_n$ are eigenfunctions of the harmonic oscillator (cf. Eq. (6-88)) and are therefore orthogonal. Consequently, since $\mu_0$ is a constant, the first integral is zero. We can conclude then that a constant dipole moment is not sufficient for a vibrational transition to occur. In order for a diatomic molecule to undergo a vibrational transition, the second integral must be non-zero, i.e., the molecule's dipole moment must change during a vibrational cycle. Homonuclear diatomic molecules, like $H_2$ and $Cl_2$, therefore have no infra-red spectra.

The simplest way to consider the energy changes that accompany vibration-rotation transitions is to consult the diagram in Fig. 14-5. As in the case of pure rotational spectra, restrictions resulting from consideration of the dipole moment are not the only restrictions that operate. Again, we shall not prove the selection rules; however, vibrational transitions for the harmonic oscillator occur only if

$$\Delta v = \pm 1 \tag{14-50}$$

where $v$ is the vibrational quantum number that occurs in Eq. (6-92). For absorption, the appropriate part of this selection rule is $\Delta v = +1$. The selection rule for the rotational changes for most molecules is still $\Delta J = \pm 1$, and in this case both possibilities are operative in absorption spectroscopy. Taking these two selection rules into account, we can draw the allowed transitions as shown in Fig. 14-5.

To determine expressions for the energy changes, we consider the allowed energies $E_v$ of the harmonic oscillator (cf. Eq. (6-92)):

$$E_v = (v + \tfrac{1}{2})h\nu_0 \qquad v = 0, 1, 2, 3, \ldots \tag{14-51}$$

Here $\nu_0$ is the *fundamental vibration frequency* of the oscillator and is given by

$$\nu_0 = \frac{1}{2\pi}\left(\frac{k}{\mu_m}\right)^{1/2}$$

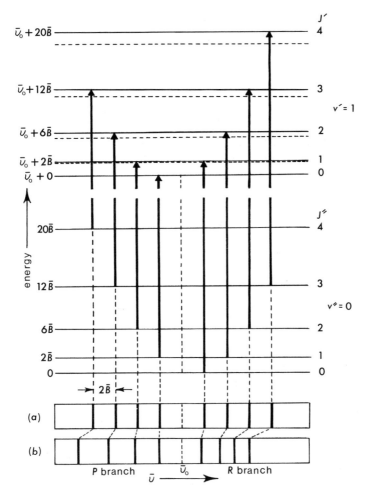

**Fig. 14-5** Schematic representation of vibration-rotation transitions. (a) is the spectrum obtained if vibration-rotation coupling is neglected; (b) is the spectrum obtained if vibration-rotation coupling is taken into account. For clarity, the transitions that produce spectrum (b) are not shown.

(We have written $\mu_m$ for the reduced mass to distinguish it from the symbol we have been using for the dipole moment.) To the energy in Eq. (14-51) we must add the rotational energy, as given by Eq. (6-46)—the result is an expression for the total vibrational and rotational energies:

$$E_{vJ} = \left(v + \frac{1}{2}\right)h\nu_0 + J(J+1)\frac{\hbar^2}{2I} \qquad (14\text{-}52)$$

Let us first consider those transitions for which $\Delta v = +1$ and $\Delta J = +1$. For the transition $v = 0$ to $v = 1$ we get

$$E_{v'J'} - E_{v''J''} = \frac{3}{2}h\nu_0 + (J''+1)(J''+2)\frac{\hbar^2}{2I} - \frac{1}{2}h\nu_0 - J''(J''+1)\frac{\hbar^2}{2I}$$

$$= h\nu_0 + 2(J''+1)\frac{\hbar^2}{2I} \qquad J'' = 0, 1, 2, 3, \ldots \qquad (14\text{-}53)$$

Here $J''$ is the rotational quantum number for the lowest rotational state. Following the same procedure we did when considering pure rotational spectra, we can determine the frequency of the quanta of radiation required to bring about transitions; it is

$$\nu = \nu_0 + 2(J''+1)\frac{h}{8\pi^2 I} \qquad (14\text{-}54)$$

or, expressing this result in wave-number units,

$$\bar{\nu} = \bar{\nu}_0 + 2(J''+1)\bar{B} \qquad J'' = 0, 1, 2, 3, \ldots \qquad (14\text{-}55)$$

Similarly, for the transitions for which $\Delta v = +1$ and $\Delta J = -1$ we get

$$\bar{\nu} = \bar{\nu}_0 - 2J''\bar{B} \qquad J'' = 1, 2, 3, \ldots \qquad (14\text{-}56)$$

Thus we see that the allowed transitions, as shown in Fig. 14-5, lead to a spectral band that is made up of a series of lines with a gap in the center corresponding to the missing $\Delta J = 0$ transitions (all of which would correspond to $\nu_0$). Figure 14-5 indicates that the central gap in the band can be utilized to deduce the spacing between the vibrational levels of the molecule. The spacing between the lines in the band is $2\bar{B}$. It is therefore possible to use vibration-rotation spectra to calculate moments of inertia and bond lengths, as was done with pure rotational spectra.

It is normal to call the set of lines at the low frequency end of the vibration-rotation spectrum—i.e., that resulting from $\Delta J = -1$ transitions—the *P branch*; the set at the high frequency end—i.e., that resulting from the $\Delta J = +1$ transitions—is called the *R branch*. (For some molecules the $\Delta J = 0$ transitions are allowed—this results in the appearance of a *Q branch* in the spectrum.)

When discussing vibration-rotation spectra we have used the rigid rotator—harmonic oscillator approximation. When we considered pure rotational spectra we mentioned that the rigid rotator model is not an entirely accurate one and that if we wished to be more exact, we would have to allow for non-rigidity of the molecule. To obtain a more accurate theory of vibration-rotation spectroscopy we would have to take into consideration the non-rigidity of the molecule. We would also need to recognize that the molecule is not strictly a harmonic oscillator. Referring back to Section 6-4, we can recall that the energy levels for the harmonic oscillator are evenly spaced. The anharmonicity that occurs in real molecules causes the vibrational energy levels to be unevenly spaced. This means that the transition frequencies will be functions of the particular vibrational states involved in the transitions. A second effect of anharmonicity is that it modifies the wavefunctions so that the selection rules have to be modified. Thus for real molecules it is possible for transitions to occur that correspond to $\Delta v = \pm 2, \pm 3$, etc.; however, these are weak compared to the $\Delta v = \pm 1$ transitions.

A further assumption we have made when considering vibration-rotation spectra is that $\bar{B}$ is constant for different vibrational states of the molecule. In practice this is not true. $\bar{B}$ includes the moment of inertia, which in turn is dependent on the internuclear distance. The average internuclear distance in an excited vibrational state of the molecule is in general larger than in the ground state. Therefore, $\bar{B}$ is a function of the vibrational quantum number $v$. This effect is known as *vibration-rotation coupling*. We can attempt to correct for it by writing $\bar{B}$ in the form

$$\bar{B}_v = \bar{B}_0 - \alpha(v + \tfrac{1}{2}) \tag{14-57}$$

where we have now written $\bar{B}_v$ to show specifically that $\bar{B}$ is a function of the vibrational quantum number; $\bar{B}_0$ is the rotational constant for the ground vibrational state; $\alpha$ is a constant that allows for the vibration-rotation coupling. The effect of vibration-rotation coupling on the vibration-rotation spectrum is to make the lines in the $R$ branch progressively more closely spaced and the lines in the $P$ branch progressively further apart as $J$ increases — an indication of this modification in the spectrum is given in Fig. 14-5.

We have given here only the very briefest consideration to the corrections that have to be made to our basic theory when we treat real molecules. To pursue the matter further would be beyond the scope of this text. However, if you wish further details, you can consult, for example, reference F1 in the Bibliography.

## 14-7  ELECTRONIC SPECTRA

In general, the energy differences between electronic states of a molecule are much larger than the differences between vibrational states of a single electronic state. The transitions that produce electronic spectra are therefore observed in the visible or ultra-violet regions of the electro-magnetic spectrum. When these electronic transitions take place we would expect changes in the vibrational and rotational states also. The result is spectra that are of considerable complexity. Here we shall dis-cuss only the most general features of the electronic spectra of diatomic molecules. We shall not discuss the details of the vibrational and rotation-al changes that accompany electronic transitions, but we should mention that, in general, electronic spectra provide more information about the vibrational and rotational structure of molecules than the vibration-rota-tion spectra do. For example, the infra-red vibration-rotation spectrum of a molecule gives information only about the ground state, whereas the electronic spectrum gives information about excited states too.

In Fig. 14-6 we show the potential energy curves for two different

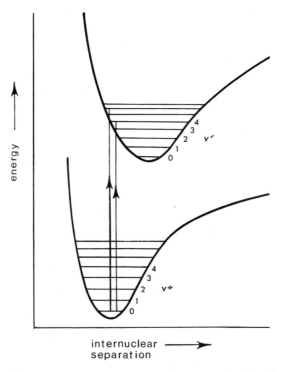

**Fig. 14-6** Potential energy curves for two electronic states of a diatomic molecule with some probable spectroscopic transitions. Note that the transitions start near the center of the $v = 0$ state and finish near the ends of the vibrational levels in the upper state.

electronic states of a molecule. Superimposed on the electronic states are the first few vibrational levels associated with them. The diagram shows the ground-state curve and an excited-state curve. In general, there will be other excited state curves, but we have shown only one for simplicity.

We now consider a transition from the ground state to the excited state. In doing so, we have to ask: Which transitions can occur between the vibrational states of the ground electronic state (labeled with $v''$) and the vibrational states of the excited electronic state (labeled with $v'$)? For vibrational transitions within an electronic state we recall that the selection rule $\Delta v = \pm 1$ is operative. For electronic transitions there are no such restrictions.

Though there is no selection rule of the type we encountered when examining vibration-rotation transitions, we can predict which transitions between the vibrational states of the two electronic states will be most probable by using the so-called *Franck-Condon principle*. This principle is a result of the fact that electronic transitions in a molecule take place so rapidly compared to the vibrational motion that there is no significant change in the internuclear distance during a transition. Taking this principle into account, we can understand why the lines drawn in Fig. 14-6 to represent transitions are vertical.

There is another factor we have to consider when trying to predict which transitions will be most probable. For any particular vibrational state of a molecule, the probability of the molecule having a particular internuclear distance is a function of the internuclear distance. We can expect the most probable transitions to be those that connect two states of high probability. In Fig. 14-6 we show some probable transitions — these have been drawn on the basis of the facts that the most probable internuclear distance for a $v = 0$ state is the center of the state whereas the most probable internuclear distances for other states are near the ends of the lines representing the states.

In this brief examination of electronic spectra we have said nothing about rotational energy changes. In fact, there are rotational states associated with each vibrational state, the consequence being that vibrational-electronic transitions are accompanied by rotational energy changes. Each vibrational component of an electronic band spectrum therefore has considerable fine structure resulting from the rotational energy changes.

In a book of this sort we have not been able to examine the rotational complexities of electronic spectra. In fact, we have been able to given only the briefest discussion of these spectra. We have confined ourselves to discussion of absorption spectra and have not mentioned emission spectra at all. Furthermore, we have not examined polyatomic molecules; nor have we examined such phenomena as fluorescence and phosphorescence. For more details of these and other aspects of molecular spectroscopy you can consult, for example, reference F1 in the Bibliography.

## SUMMARY

1. In this chapter we examined the very fundamental theory of time-dependent phenomena, using spectroscopic theory as an example.

2. We considered a system with just two stationary states and examined the effect of a time-dependent perturbation on the system. If we label the state with lower energy $m$ and the state with higher energy $n$, the equation that gives the rate of transition to state $n$ is

$$\frac{da_n}{dt} = -\frac{i}{\hbar} \exp\left(\frac{i(E_n - E_m)t}{\hbar}\right) H'_{nm} \tag{14-10}$$

where $a_n$ is related to the probability of finding the system in the upper state.

3. We then considered the specific case where the perturbation is due to electromagnetic radiation. The equation we got for the rate of transition from a lower state $m$ to an upper state $n$ was

$$\frac{d(a_n^* a_n)}{dt} = B_{mn}\rho \tag{14-28}$$

where $\rho$ is the radiation energy density and $B_{mn}$ is the *Einstein coefficient for induced absorption*; $B_{mn}$ includes the *transition moment* $|\mu_{mn}|$.

4. We related the theory to experimental measurements by relating $B_{mn}$ to the experimentally determined *absorption coefficient*.

5. To examine some of the important applications of the theory in experimental work we very briefly considered molecular spectroscopy for diatomic molecules.

6. *Pure rotational spectra* result from transitions between the allowed rotational states of molecules. Using the rigid rotator approximation, we saw that a diatomic molecule can produce pure rotational spectra only if it has a non-zero permanent dipole moment. The spacing between the lines in a pure rotational spectrum (in the rigid rotator approximation)

is $2\bar{B}$ where

$$\bar{B} = \frac{h}{8\pi^2 I c} \qquad (14\text{-}45)$$

$I$ is the moment of inertia of the molecule.

7. *Vibration-rotation spectra* occur when molecules undergo transitions between vibrational states and their associated rotational states. Using the harmonic oscillator approximation for a diatomic molecule, we saw that these spectra can be produced only by molecules that possess dipole moments that change during the vibrational cycle. For most diatomic molecules the vibration-rotation spectrum resulting from a transition from one vibrational state to another is a band consisting of a series of lines (with spacing $2\bar{B}$) with a gap in the center.

8. We discussed *electronic spectra* of diatomic molecules very briefly by considering the possible vibrational changes that accompany a transition between two electronic states. We saw that although no selection rule operates for the vibrational changes, the *Franck-Condon principle* and the probabilities for the molecule having particular internuclear distances allow us to predict the most probable transitions.

## EXERCISES

**14-1**   Generalize the theory of Section 14-2 so that it can be applied to systems with more than two stationary states; i.e., verify Eq. (14-11).

**14-2**   Referring to the discussion of Chapter 3 and to Fig. 3-2, consider an electron in a one-dimensional box of length $a$. Further, consider that the electron's charge is balanced by a positive charge located at the mid-point of the potential well. If the electron is originally in the state for which $n_x = l$ and undergoes a transition to the state for which $n_x = m$, show that the transition moment is given by

$$|\mu_{ml}| = \frac{ea}{\pi^2}\left[\frac{\cos\,(l-m)\pi - 1}{(l-m)^2} - \frac{\cos\,(l+m)\pi - 1}{(l+m)^2}\right]$$

**14-3**   (a) Using the result of Exercise 14-2, deduce the selection rules for transitions between the first four states of the system; i.e., which transitions between these states are allowed, and which are not allowed? (b) These same selection rules can be determined very simply by arguments based on the symmetry of the system. Consider the general form of the wavefunctions as given by Eq. (3-19) and illustrated in Fig. 3-2, place the origin for the coordinate system at the center of the potential, and examine the symmetry of the transition moment integrals. By doing this, obtain the same result as you did in part (a).

**14-4**  Carbon monoxide absorbs energy in the microwave region of the electromagnetic spectrum at a frequency of $1.15 \times 10^5$ Mc sec$^{-1}$. If this absorption is due to a transition between the rotational levels for which $J = 0$ and $J = 1$, calculate the moment of inertia and equilibrium internuclear distance for the molecule.

**14-5**  Carbon monoxide exhibits an intense spectral band at $2144$ cm$^{-1}$ (i.e., in the near infra-red region of the electromagnetic spectrum.) Use this knowledge to calculate; (a) the fundamental vibration frequency of the molecule, (b) the period of vibration, (c) the force constant.

# References

1. Schroedinger, E., *Ann. Physik*, **79**, 361, 489, 734 (1926); **81**, 109 (1926).
2. Born, M., *Z. Physik*, **37**, 863 (1926); **38**, 803 (1926).
3. Heisenberg, W., *Z. Physik*, **43**, 172 (1927).
4. Pauli, W., *Z. Physik*, **43**, 601 (1927).
5. Pauli, W., *Z. Physik*, **31**, 765 (1925).
6. Hartree, D. R., *Proc. Cambridge Phil. Soc.*, **24**, 89, 111 (1928).
7. Fock, V., *Z. Physik*, **61**, 126 (1930).
8. Slater, J. C., *Phys. Rev.*, **35**, 210 (1930).
9. Slater, J. C., *Phys. Rev.*, **36**, 57 (1930).
10. Born, M., and Oppenheimer, J. R., *Ann. Physik*, **84**, 457 (1927).
11. Wind, H., *J. Chem. Phys.*, **42**, 2371 (1965).
12. Hellmann, H., *Einfuhrung in die Quantumchemie*, Franz Deuticke, Leipzig, 1937, p. 133.
13. Heitler, W., and London, F., *Z. Physik*, **44**, 455 (1927).
14. Slater, J. C., *Phys. Rev.*, **37**, 481 (1931); **38**, 1109 (1931).
15. Pauling, L., *J. Am. Chem. Soc.*, **53**, 1367, 3225 (1931); **54**, 988, 3570 (1932).
16. Weinbaum, S., *J. Chem. Phys.*, **1**, 593 (1933).
17. Rosen, N. *Phys. Rev.*, **38**, 2099 (1931).
18. Kolos, W., and Roothaan, C. J., *Rev. Mod. Phys.*, **32**, 219 (1960).
19. Kolos, W., and Wolniewicz, L., *J. Chem. Phys.*, **41**, 3663 (1964).
20. Hückel, E., *Z. Physik*, **70**, 204 (1931); **76**, 628 (1932).

# Bibliography

The following is a list of reference material and suggested additional reading. Many of the books listed give references to the original work in quantum theory and quantum chemistry—reference C2 is especially good in this respect.

**A.** The following books give non-mathematical introductions to quantum mechanics and bonding theory:

1. Coulson, C., *Valence*, Oxford University Press, London, 1961.
2. Pauling, L., *The Nature of the Chemical Bond*, 3rd Edition, Cornell University Press, Ithaca, New York, 1960.

**B.** The following books cover the same material as this book, and at approximately the same level. Many of them give a somewhat expanded coverage and go more deeply into applications of the basic theory:

1. Anderson, J. M., *Introduction to Quantum Chemistry*, Benjamin, New York, 1969.
2. Daudel, R., *The Fundamentals of Theoretical Chemistry*. Translated from the French by R. E. Watts, Pergamon, London, 1968.
3. Hanna, M. W., *Quantum Mechanics in Chemistry*, 2nd Edition, Benjamin, New York, 1969.
4. Kauzmann, W., *Quantum Chemistry*, Academic Press, New York, 1957.
5. La Paglia, S. R., *Introductory Quantum Chemistry*, Harper and Row, New York, 1971.
6. Murrell, J. N., Kettle, S. F. A., and Tedder, J. M., *Valence Theory*, Wiley, New York, 1965.
7. Pauling, L. and Wilson, E. B., *Introduction to Quantum Mechanics*, McGraw-Hill, New York, 1935.

**C.** The following books cover the material in this book, but at a more advanced level:

1. Eyring, H., Walter, J., and Kimball, G. E., *Quantum Chemistry*, Wiley, New York, 1944.
2. Pilar, F. L., *Elementary Quantum Chemistry*, McGraw-Hill, New York, 1968.

**D.** The following is a selection from the many books on quantum theory that have been written more for physicists than for chemists—most of them treat the subject in more mathematical detail than is usually found in books on quantum chemistry:

1. Bates, D. R., Ed., *Quantum Theory*, Volume 1, Academic Press, New York, 1961.
2. Bohm, D., *Quantum Theory*, Prentice-Hall, Englewood Cliffs, N.J., 1951.

3. Dirac, P. A. M., *The Principles of Quantum Mechanics*, 4th Edition, Oxford University Press, London, 1958.
4. Hameka, H. F., *Introduction to Quantum Theory*, Harper and Row, New York, 1967.
5. Landau, L. D., and Lifshitz, E. M., *Quantum Mechanics: Nonrelativistic Theory*, Translated from the Russian by J. B. Sykes and J. S. Bell, Addison-Wesley, Reading, Mass., 1958.
6. Rojansky, V., *Introductory Quantum Mechanics*, Prentice-Hall, Englewood Cliffs, N.J., 1938.

**E.** The following books cover all the mathematics needed for this book:
1. Boas, M. L., *Mathematical Methods in the Physical Sciences*, Wiley, New York, 1966.
2. Margenau, H., and Murphy, G. M., *The Mathematics of Physics and Chemistry*, Van Nostrand, Princeton, 1956.

**F.** The foregoing books will serve as reference material for Chapters 1 to 7 and Chapters 9 to 11. Some additional reference material for the other chapters is listed below:

*Chapter 8*
There are chapters on the application of group theory to chemistry in, for example, references B5, C1, and C2. The following references cover the introductory theory and applications. The first is concerned only with applications in spectroscopy: the others treat applications in spectroscopy and in other fields of chemistry:
1. Barrow, G. M., *Introduction to Molecular Spectroscopy*, McGraw-Hill, 1962.
2. Cotton, F. A., *Chemical Applications of Group Theory*, 2nd Edition, Wiley, New York, 1971.
3. Orchin, M. M., and Jaffe, H. H., *J. Chem. Educ.*, **47**, 246, 372, 510 (1970).

*Chapter 12*
The material of this chapter is treated in more detail than here in, for example, references B5 and C2. The following books deal exclusively with the application of MO theory to organic molecules:
4. Salem, L., *The Molecular Orbital Theory of Conjugated Systems*, Benjamin, New York, 1966.
5. Streitwieser, A., *Molecular Orbital Theory for Organic Chemists*, Wiley, New York, 1961.

*Chapter 13*
There are chapters on the material of this chapter in references B5 and F2. The following are also useful references:
6. Ballhausen, C. J., *Introduction to Ligand Field Theory*, McGraw-Hill, New York, 1962.
7. Orgel, L. E., *An Introduction to Transition Metal Chemistry*, 2nd Edition, Methuen, London, 1966.
8. Cotton, F. A., *J. Chem. Educ.*, **41**, 466 (1964).

*Chapter 14*
Reference F1 gives a very good introduction to the theory of molecular spectroscopy. Other excellent books are the following:
9. Herzberg, G., *Atomic Spectra and Atomic Structure*, Dover, New York, 1944.

10. Herzberg, G., *Spectra of Diatomic Molecules,* Van Nostrand, Princeton, N. J., 1950.
11. Herzberg, G., *Infrared and Raman Spectra*, Van Nostrand, Princeton, N.J., 1945.
12. Sandorfy, C., *Electronic Spectra and Quantum Chemistry,* Prentice-Hall, Englewood Cliffs, N.J., 1964.
13. Wilson, E. B., Jr., Decius, J. C., and Cross, P. C., *Molecular Vibrations*, McGraw-Hill, New York, 1955.

# Index